Were These
King Solomon's Mines?

NEW ASPECTS OF ARCHAEOLOGY

Edited by Sir Mortimer Wheeler

Were These King Solomon's Mines?

Excavations in the Timna Valley

Beno Rothenberg

25 Color Plates
128 Monochrome Plates
77 Line Drawings

STEIN AND DAY,
PUBLISHERS, NEW YORK

To
Samuel and Milada Ayrton,
Patrons of Timna,
in sincere appreciation and friendship

First published in the United States in 1972

Library of Congress Catalog No. 72–82213

Printed in Great Britain by Hazell Watson and Viney Ltd, Aylesbury
Stein and Day, Publishers, *7 East 48 Street, New York, N.Y. 10017*

ISBN *0–8128–1506–8*

Contents

General Editor's Preface

In 1959 an expedition, which eventually assumed an international complexion, began under the leadership of Dr Beno Rothenberg the exploration of the region lying along the western flank of the Wadi Arabah between the southern end of the Dead Sea and the head of the Gulf of Elat-Aqaba. The region might without excess be described as the abomination of desolation. The great Wadi and its feeders conduct the liquid mud churned up by the brief winter-rains without mitigating substantially the obstinacy of the arid soil. Mostly the climate burns with a sultry heat during the day and bites icily at night (when I have thereabouts scooped hollows in the sand to break the blast). This is no place for pause or settlement of a casual kind. But there is one factor which for six thousand years has drawn men intermittently to the neighbourhood of the great sandstone cliffs, and particularly those of Timna, that confront the lands of the Wadi and mask them from the westward deserts of the Negev and Sinai. That factor is the copper ore which occurs in desirable abundance along the foot of those cliffs and has encouraged exploitation by Arabians, Egyptians, Romans, and their medieval and modern successors. Nowhere else can be seen so extensive a range, whether in time or in space, of the technologies of copper-mining through the ages.

It is thus a land of mining-camps that forms the rugged but picturesque background of Dr Rothenberg's narrative. He justly describes this as the first firm archaeological foundation for the history of copper-production in the ancient world. Before 3000 BC North Arabians were already supplementing the collection of surface-material with the use of heavy stone axes, and were smelting the ore in primitive furnaces or bowl-hearths. Between the fourteenth and the twelfth centuries BC the Ramesside pharaohs employed here an organized local labour-force, which has left a vivid display of rock-drawings showing scenes of hunting and fighting, with archers (presumably Egyptians) in chariots. Its smelting-furnaces were structurally improved upon those of their predecessors, with provision for manipulating bellows and shielding them from the intense heat generated during the process. Not least, in the latter years of the fourteenth century BC, the camps included an Egyptian shrine built, and later rebuilt, in the name of that universal provider, the goddess Hathor. Understandably the shrine was of no great pretension, but at least it yielded almost 10,000 small finds, amongst them inscribed offerings which have furnished a useful guide to the history of the place.

Towards the end of the twelfth century BC the rebuilt temple was destroyed, apparently by an earthquake, and about the same time Egyptian control was replaced by that of Midianites from the Hedjaz. Vestiges of Hathor were now consigned to the rubbish-tips, the ruined shrine was roofed with a tent presumably as a desert sanctuary, and its holy of holies received a Midianite copper snake with a gilded head as its only votive object.

In spite of traditional associations of King Solomon with the mines and the landscape, the great king is probably the most eminent absentee from the archaeological sequence. But the copper-field was re-opened in the second century AD on an organized scale by experienced Roman engineers when the Third Legion *Cyrenaica* occupied the Arabah after the displacement of the Nabataean régime. As in their iron-mining, the Romans armed with their metal chisels cut shafts deeply into the cliff-sides where copper ore was found to be particularly profitable. Only rough dressing of the embedded metal took place in the immediate vicinity, where there are large heaps of broken sandstone. For smelting, the material thus roughly prepared was concentrated at a unitary smelting-centre near water and timber some distance to the south, beside the Roman highway from Palestine to the Gulf. The furnaces used at the centre were generally similar to those of the Egyptian period, and indeed comparable devices together with crucible-furnaces of Roman type for the casting of ingots or implements continued unchanged into the medieval period.

During the past twenty years modern methods have been imported into the Timna area and the ore-deposits are again being busily exploited. In the interests of archaeological and historical record, the intervention of Dr Rothenberg and his team was a timely one, and, as the following pages will show, it has added a new and instructive chapter to the history of technology.

MORTIMER WHEELER

Introduction

Timna, the valley of the copper smelters, was never an isolated desert island, the scene merely of exploitation of natural wealth, although this was, obviously, an important factor and part of far reaching historical events. Geographically, it is located about 30 km. north of Elat, on the west side of the southern Arabah. The Arabah, the geographical background of the Timna story, is part of the large Palestinian Rift, which contains in the north the river Jordan, the Sea of Galilee and, further south, the Dead Sea. Starting at the southern end of the Dead Sea, the Arabah ends at the shore of the eastern arm of the Red Sea, known as the Gulf of Elat-Aqaba. This 175 km. long and extremely dry strip of land separates the high plateau of eastern Palestine (Jordan) and the Hedjaz from the southern Negev and adjacent Sinai. An eternal no-man's land, it lies at the crossroads between two continents – Asia and Africa – has a border of four countries – Israel, Jordan, Arabia and Sinai – and is one of the major centres of Biblical history. Here decisive events took place in the early history of the Jewish people and their Kenite-Midianite relations as well as in the history of Pharaonic Egypt, of Edom, the Nabataeans and the Roman Empire.

Plate 1

The Timna Valley and related areas of the Arabah, where copper was mined in ancient days, have lately acquired a new historical significance. The recent archaeological survey of the Arabah and the adjacent Sinai wilderness has shown that the copper mining areas and their environments were not the sole objectives of temporary foreign occupation and exploitation but were of vital importance and often major incentives for the history of the ancient inhabitants, indigenous or intrusive, of the Arabah and all adjoining lands.

Fig. 1

Over 650 ancient sites, previously unknown, were recently discovered in the large arid expanse of the southern Arabah, the southernmost Negev and Sinai, from the Suez Canal to the Mountains of Edom (Jordan). Although lack of space does not permit here a detailed report, which will be published in a forthcoming volume on the Sinai peninsula, a general description of the most relevant survey results is indicated here. 'Sinai' must be considered

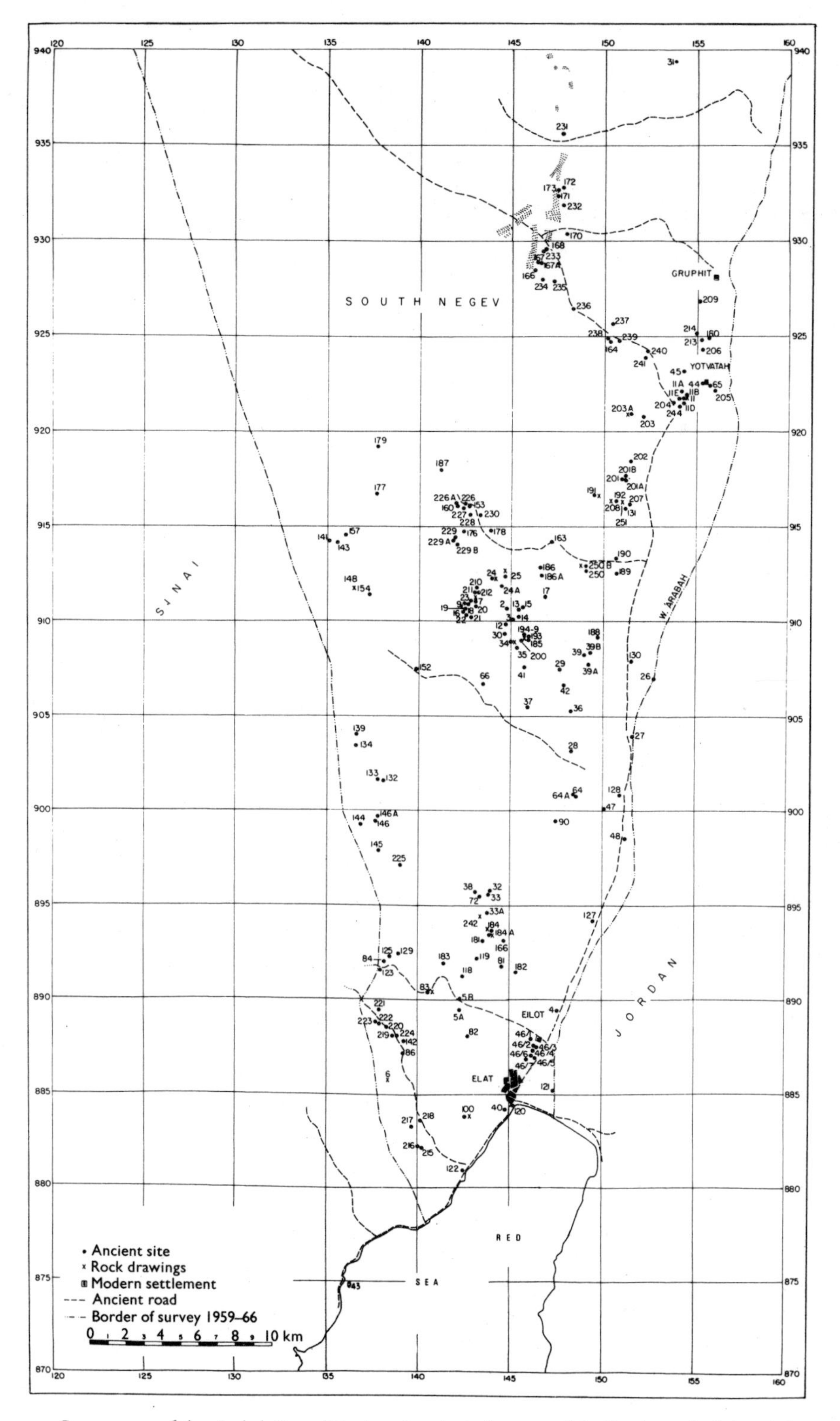

1 *Survey map of the Arabah Expedition's archaeological survey of the Southern Arabah and the southernmost Negev.*

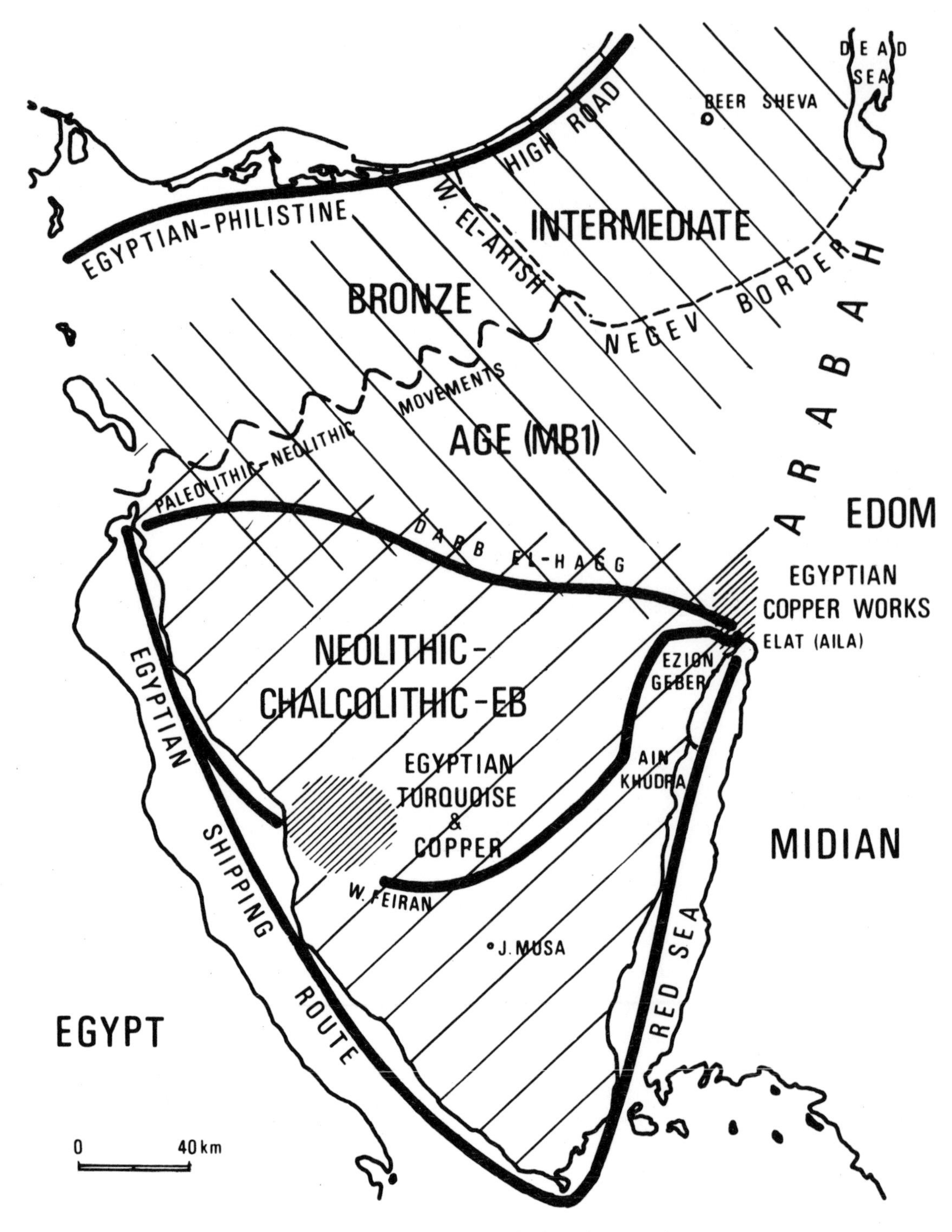

2 *Map of Sinai indicating the results of recent explorations by the author's expedition*

as a collective name for three completely different landscapes, essentially different also as regards their history: the Mediterranean coastal strip is the main thoroughfare between Asia and Africa, a highroad throughout most of the history of Palestine and Egypt, as is clearly shown by numerous ancient sites to be found all along its route. The northern and Central Sinai desert, which is at present being surveyed by the Arabah Expedition, begins actually only some 10–20 km. south of the Mediterranean coast and rises slowly, where it becomes the high plateau of the Badiyat el Tih, the 'Wilderness of Wandering', a terrible, hot, dry and treeless wasteland. Yet this forbidding wilderness was not completely empty. Although no sedentary civilization seems to have existed here at any time, Middle Palaeolithic and Pre-pottery Neolithic B camping sites were found indicating lines of movement between the narrow landbridge connecting the northernmost tip of the Gulf of Suez with the Little Bitter Lake and the mountains of the Central Negev, particularly the area of Kadesh Barnea. The greatest surprise, however, was the discovery throughout the width of Central Sinai, from the Negev to the very bank of the Suez Canal, of numerous groups of habitation sites from the Intermediate Bronze Age (also referred to as the Early Bronze-Middle Bronze Age or Middle Bronze Age I), and dated to the last centuries of the third millennium BC. There can be no doubt that at this period there were intensive movements of apparently warlike nomadic tribes across the Great Wilderness, from Palestine into Egypt.

Fig. 2

It was no less surprising to see that these Intermediate Bronze Age camping sites, temporary settlements and burials did not continue south of a line drawn roughly between Elat and Suez, in spite of the fact that a metal-using, warlike people should have been most interested in the mineral deposits of southern Sinai and the southern Arabah.

Southern Sinai, with its almost 250 newly discovered ancient sites, presents a completely different historical picture, related directly to the Timna excavations. As it happened, the Timna excavations and the explorations of South Sinai and the southern Arabah are complementary investigations. The mining and habitation sites of Sinai and the Arabah presented the geographical and archaeological background necessary for a real understanding of the Timna excavations, without which Timna would remain historically an enigmatic island in what was hitherto considered a vast, seemingly empty, desert. On the other hand, without the Timna excavations Sinai and the Arabah, and also much of North Arabia and some of Egypt's history would be hard to understand and many of their ancient sites would be lacking the certain dates and the technological understanding achieved by the Arabah excavations.

Although the deep, sandy, and granite wadis of South Sinai provide evidence for Pre-pottery Neolithic B intrusions, possibly for the purpose of extracting turquoise, the most striking fact about the archaeology of southern Sinai is its dense occupation during the Chalcolithic and Early Bronze Age periods in the fourth and third millennia BC. During these periods copper-hungry intruders from Arabia, Egypt and Palestine made this area a vital landbridge between Asia and Africa. In these early, prehistoric times, as again in the second millennium BC and in the Roman period, the great desert track from the Delta to the mountains of Edom and Midian, across and including the Arabah, must be considered as one historico-geographical unit. Timna was then but an important station on the bridgehead, with many more Chalcolithic and Early Bronze Age copper mining and smelting sites all over South Sinai and the southern Arabah as functional meeting places between East and West. It was also here that the Egyptian pharaohs of the Old Kingdom met their 'Asiatic' enemies, who were obviously the Early Bronze Age II miners and their people, infiltrating into South Sinai from Canaanite Early Bronze Age II Palestine.

When, almost 2000 years later, the Pharaohs of the New Kingdom penetrated and colonized Palestine, and probably also Edom, Timna became a major copper producing site. In fact, the Ramesside kings activated only two major areas in the vast desert tract between the Delta, Palestine and Edom: the turquoise mines of south-western Sinai and the copper mines of the Arabah. The rest of this expanse seems not to have had any attraction for the Egyptian kings and it was left to the not always friendly nomads of the desert. With the decline of Egypt's power in the twelfth century BC the Egyptian copper mining expeditions no longer reached Timna and, once more, the scene was left to the local tribes.

Almost 1500 years later Roman Imperial forces occupied the southern Arabah and also took over the ancient roads into Sinai. The Third Legion *Cyrenaica* seems to have restarted the copper industry of the area, in the Arabah as well as in South Sinai. The ancient track from the northern shore of the Red Sea towards South Sinai, previously used by waves of Neolithic and Chalcolithic settlers, miners and metallurgists, became a Roman road connecting Aila on the Red Sea with the Roman station of Pharan in the Wadi Feiran. This, once more, was a period of intensive activity in the copper mines of the area and Timna, Beer Ora and other sites in the Arabah, were part of the far-flung activities of the Romans. The decline of the Roman Imperium, the disorder and reform of the Byzantine regime and the Arab conquest, all left their traces in the area and in Timna, but these were vague traces of only limited, local significance. Timna had to wait for almost 2000 years until Israel's Timna Copper Mines Ltd. turned the area once more into a centre of intensive copper production.

The Arabah Expedition, directed since its inception in 1959 by the author, recently concluded its first series of systematic excavations on metallurgical sites in the Timna Valley and the Arabah belonging to historical periods covering almost 6000 years. Due to lacunae in the exploitation of the mines, the different organization involved, and the abandonment of the Timna Valley for almost 2000 years until recent times, many ancient installations were found well preserved. Their discovery and excavation resulted in a new, archaeologically well-documented history of the technology of copper. As a result of archaeologically favourable circumstances Timna seems to be the only place known today where such ancient and continuous copper working traditions are found to be represented by extensive production plants. Here, 6000 years of copper technology can be followed from the actual, datable copper mines to the finished copper objects. Yet these unusual favourable circumstances would have gone scientifically unexploited had it not been for the tireless efforts of numerous people and institutions whose collaboration made this lengthy and involved undertaking possible. It is with great appreciation and humble thanks that the author acknowledges the important contributions by the members of the Arabah Expedition's research team: Professor Y. Aharoni, head of the Archaeological Institute at Tel Aviv University; Professor A. Lupu, (Extractive Metallurgy) Israel Institute of Technology, Technion, Haifa; Dr R. F. Tylecote, (Metallurgy and Metallography) Newcastle University; and Professor P. Fields, (Chemistry) Argonne National Laboratory. This team has recently been joined by Dr H. G. Bachmann (Chemistry), Frankfurt. Many thanks for invaluable advice during many years are due to the late Bentley H. McLeod, our first metallurgical advisor at Timna; Dr H. H. Coghlan and Professor R. J. Forbes.

Essential and responsible work in the field was done by numerous volunteers from many countries. Detailed lists and acknowledgements will be published in the final excavation reports, but we should especially mention here with many thanks the devoted work of Ezra Cohen, Morag and Paul Woudhuysen, Haya Kaminsker, Joram Lemberg, Kirsten Logstrup, Barbara Hall, Ron Elul, Alfons Nussbaumer, Brian O'Dea, Rochele Zaltzman, Craig Meredith, C. W. Brewer and H. Arthur Bankoff. The author will always remember the courageous behaviour of a group of young American student volunteers who reported to work at our base in Elat at the usual early hour of a hot summer day in 1969, in spite of the fact that Elat was being shelled at the very time from across the Jordanian border.

Lack of space does not permit the expression of our thanks here individually to all of the many scientists, area supervisors, and technicians who assisted us in various ways during the eleven years of work of the Arabah Expedition. However, of special importance was the contribution in the field of the archaeologists Miss E. Yeivin (1965), Professor S. Applebaum (1964) and of the geologists Professor J. Bentor, J. Bartura and M. Price. Excavation plans were prepared by Zvi Askenazi (Haifa), Sven Thorsen, Jorgen Ilkjaer and Jorn Bie (Denmark), Shmuel Moskovits (Tel Aviv) and Reinhard Maag (Switzerland). He and Susan Maag are responsible for the final editing of all field plans and maps of the Arabah Expedition reproduced in this volume. Drawings of objects were prepared by Ruth Halfon, Naomi Schechter and Orna Semmer; photographs of objects were made by Abe Hai and Joab Marcus. Photographs of excavations and surveys were made by the author.

Many thanks are due to the author's scientific assistants at the Archaeological Institute of Tel Aviv University who conducted valuable independent research on material from the excavations, partly incorporated into the present report: Maya Zahavy, Trude Kertesz and Benett Kozloff. Specialized work on the Timna finds was done by Dr R. Giveon and Professor A. R. Schulman (Egyptology), Dr M. Gihon (Roman pottery), Giora Ilani (fauna), Daliya Hakker and Dr Lernan (bones), Dr H. Friedman (textiles), Dr N. Haas (human remains), Professor I. Slatkin (petrography of ceramics), Professor A. Fahn (botany), Gusta Lehrer (glass), Dr D. Barag (advisor on glass), Professor H. Mendelsohn (zoology). Professor Y. Aharoni was the Arabah Expedition's main advisor on stratigraphical problems and he and Miriam Aharoni made important studies of the pottery from Site 2, Timna. We also thank Dr Zvi Ron (Haifa), T. G. H. James (British Museum, London), Peter Parr (Institute of Archaeology, London), Professor A. D. Colman (Tel Aviv), Professor A. Shalit (Jerusalem), Mrs Crystal M. Bennett

(London-Jerusalem), the late Professor R. de Vaux O.P. (Jerusalem) and Peter Clayton (London) for valuable advice.

The Arabah Expedition, founded by the author in 1959 as an almost one-man archaeological field expedition, was at first aided by volunteers and private subsidy. Since 1963 the Arabah Expedition has been affiliated to the Museum Haaretz of Tel Aviv, joined in 1964 by the Technion (Professor A. Lupu) and in 1969 it became part of the Archaeological Institute of the Tel Aviv University (Professor Y. Aharoni). Its main sponsors during the last eight years were Dr K. Moosberg of Nehushtan Ltd., Tel Aviv, who also made it possible to arrange for a permanent exhibition of the Timna excavations at the Museum Haaretz, Tel Aviv; Museum Haaretz Tel Aviv and the Friends of the Museum, the Municipality of Elat, the Elot Regional Council, the Timna Copper Mines Ltd., the Tel Aviv University and the Ch. J. and F. Rothenberg Foundation.

The author's special thanks are due to Sir Mortimer Wheeler, the General Editor of New Aspects of Antiquity who initiated the writing of this volume, to Mr Peter Clayton, the archaeological editor of Thames and Hudson, for his highly qualified editorial attention to text and illustrations, and to Professor Y. Aharoni and Dr R. F. Tylecote, who read the whole manuscript and gave considerable important advice.

I

The Timna Valley

Fig. 3

The Timna Valley (formerly Wadi Mene 'iyeh), located 20–30 km. north of the Gulf of Elat-Aqaba, is a large, semi-circular, erosional formation of some 70 sq. km. On its east side, it opens towards the Arabah; to the north, west and south it is closed by almost unscalable, 500–700 m. high cliffs of dolomite and limestone, towering above colourful Nubian sandstone layers. In the centre of the Timna Valley rises Har (Mount) Timna (453 m.), formed mainly by Pre-Cambrian granites of strong red, brown and black colours. The strong contrast between the tabular dolomitic top of Har Timna and its rugged flanks and slopes, cut by numerous small but deep gullies, adds to the unique quality of this landscape.

Palaeozoic sandstone hills of varying heights and tints of red and yellow are grouped in a large half-circle around Har Timna. Weathering has created strange forms and shapes and some of these hills resemble sphinxes or large sculptures of men and monsters, others enormous pillars or altars. One of these impressive red sandstone formations, called 'King Solomon's Pillars', has recently become one of the main tourist attractions of the area but its almost mystical strangeness and beauty evidently attracted prehistoric man to camp in its shade long before.

Four wadis running from west to east, from the Timna cliffs to the Nahal Arabah, drain the Timna Valley and turn occasionally, after heavy rain, into rushing torrents that carry large masses of debris, sand and gravel. Nahal Nimra and Nahal Nehushtan pass south of Har Timna and, near the east end of the latter, the modern Timna copper mines are located. Nahal Timna, the central wadi, passes north of Har Timna with Nahal Mangan still further north, beyond Har Mikhrot. Around and on top of the Pre-Cambrian massif are found various rock formations of Palaeozoic age, of which the so-called Nehushtan and Mikhrot Formations are the mineral source of the modern Timna copper mines, mainly copper silicates, such as chrysocolla, and small quantities of atacamite, malachite, dioptase, tenorite, etc. Ancient mines are found in the Mesozoic rock formations in the lower parts of the Timna cliffs. Here, in layers called

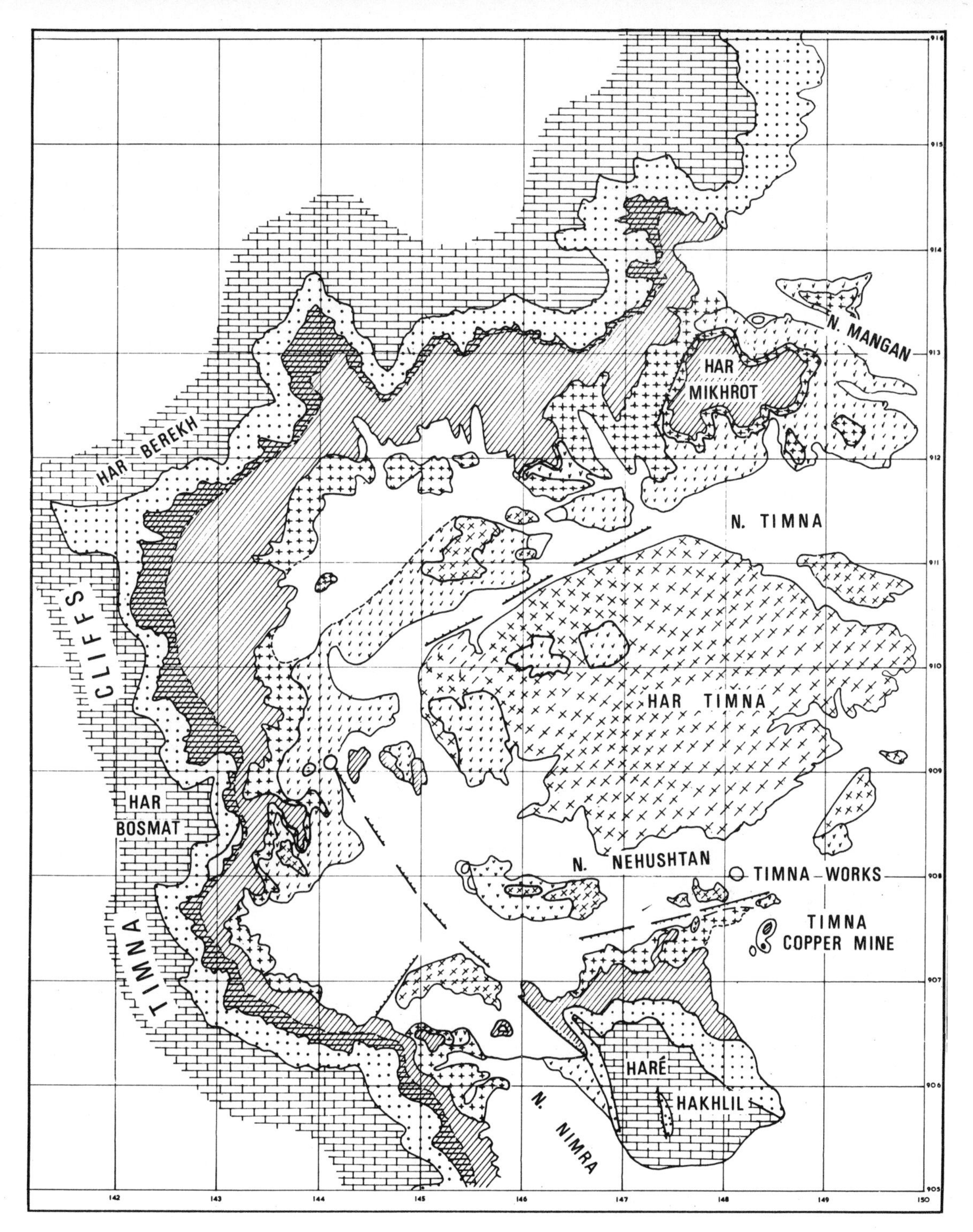

3 Geological map of the Timna Valley (by M. Price)

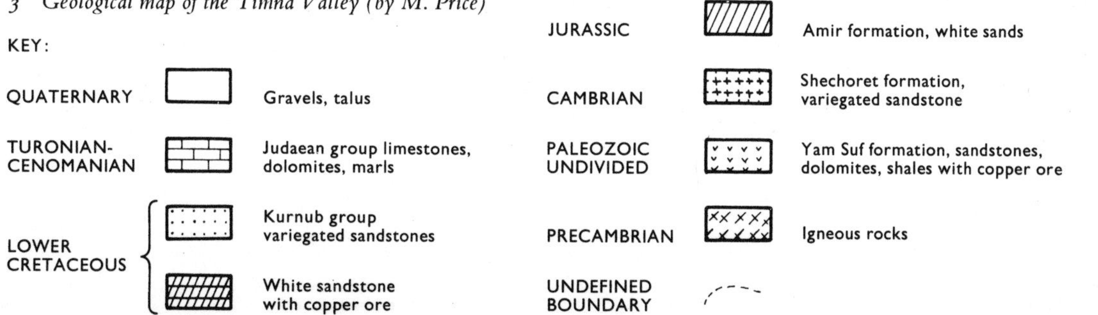

Plate VIII

by the geologists the 'Middle White' horizon, blue and grey nodules of azurite, malachite and chalcocite, containing up to 37% copper, are found in abundance.

As the Timna Valley was formed by headward erosion of the four Timna wadis, it became gradually enlarged into a huge amphitheatre or circus and, as a result of the lowering base level of the nearby Wadi Arabah, deepened from the Neocene up to Recent in several stages. Extensive remains of early levels can be found between the Timna cliffs and Har Timna as flat, high alluvial terraces, dissected by numerous recent gullies. Many of these terraces, especially those near the Middle White Sandstone layers, served as convenient locations for ancient mining camps, still *in situ*.

The Timna Valley is not the only circus formed by erosion in the south-west Arabah. Located about 12 km. north of Elat Nahal Amram (formally Wadi Amrani), like Timna, is characterized by a rugged looking magmatic mountain, Har Amram, encircled by high limestone cliffs. Here, similarly, Middle White Sandstone layers in the southern half of these semi-circular cliffs contain rich cupriferous nodules which were worked anciently. Other nodular copper-ore deposits in the western Arabah, exploited in the past, are found south of Beer Ora (formerly Bir Hindis) and in Nahal Jehoshaphat, south-west of Elat.

Timna explorations 1959–64

The Timna Valley was first mentioned as a site of copper smelting by J. Petherick in 1861 and was since partly explored and described by A. Musil (1902), F. Frank (1934) and N. Glueck (1935).

In 1959 the Arabah Expedition was organized by the author to explore the Arabah, from the Dead Sea to the shores of the Red Sea. During the course of this work the Timna Valley was inspected. Comparing the previous published information on the area with the factual data found during this first preliminary survey, it became obvious that a thorough investigation of Timna with modern scientific and improved archaeological methods would be a worthwhile undertaking.

Seven smelting camps with slag heaps had been reported in Timna by F. Frank and N. Glueck, which made it clear that somewhere in this area copper ore deposits should be found which were the ancient source for the copper smelting camps of Timna. But no actual copper mines or copper ore deposits suitable for smelting by the ancient smelting processes had ever been reported. Therefore, in January 1959 systematic explorations of the Timna Valley were started, and

Plates III, IV

the first stage of the investigations, including also Nahal Amram, was concluded in April 1961. The first two weeks were spent walking all over the Timna area and climbing the numerous deep gullies at the head of the Timna wadis with the geologist J. Bartura, a specialist on the geology of the area. Soon after the commencement of the search, at the far west end of the Timna circus, extensive copper ore deposits and evidence of large scale ancient mining activities were found. These occurred in the Middle White Sandstone horizon, right under the towering Timna cliffs. Yet the discovery of the actual ancient copper mines of Timna and Nahal Amram raised many new and rather unexpected problems. At least three different stages of mining technology could be distinguished in the mines of Timna. Many of the mining walls showed clear evidence of open cast mining, a kind of stone quarrying technique, and hand picking of stray, washed out copper ore nodules, and there was also proper horizontal and deep shaft mining. The artifacts found near the mines did not, at first, assist very much in relating the different mining techniques to different historical periods. Not only rough prehistoric flint implements and some coarse Iron Age pottery, but also Roman and Arabic sherds were found near mining sites showing different mining techniques, and these could not be correlated.

Plate V

Extensive mining camps were observed near many of the ore-bearing, white sandstone, walls. The camps were littered with ore, fluxes and stone tools, but the technological significance of the installations was by no means clear at first. The fact that only a very few sherds or flint tools were found in these camps made dating rather difficult.

Plate VI

After the initial exploration of the Middle White copper ore deposits of Timna attention was turned to the seven published smelting camps and their slag heaps; subsequently the central and eastern parts of the area were scoured for possible additional sites. To the seven camps previously reported by Frank and Glueck, were added several more new smelting sites. Here in the smelting camps a considerable quantity of pottery, much of it new material, was collected but the lack of really relevant comparative material did not allow any accurate dating. The same applied to those mining and smelting camps where flint tools, and fragments of rough pottery, indicated a prehistoric date. Perhaps the story of 'King Solomon's Mines' and the then generally accepted date of the tenth-sixth centuries BC for the Arabah mines added to the initial dating problems. It became clear that good quality metallic copper was produced in Timna from Chalcolithic times onwards and work in the field and in the laboratories concentrated on the scientific reconstruction of the processes involved in the different periods. This systematic research into the technological interconnections between the various mining and smelting installations, the tools and the principles involved in the overall organization of the metal industries of each period, made it finally possible to establish in the Arabah the existence of four distinct periods of metallurgy, *i.e.* copper mining, smelting and casting and also iron smithing.

Three major periods of early copper metallurgy and an additional, later, period of copper and iron workings were identified. In fact, complete and well established production systems, covering about 6000 years, were found commencing with the Chalcolithic period (the fourth millennium BC) operating again in the Late Bronze–Early Iron I Age, (the fourteenth to twelfth centuries BC), and again, after a long lapse, in the Roman period, to the second century AD. From Byzantine to Late Arab times several minor sites of metallurgical activities were located as well as secondary exploitation of earlier mining and smelting installations, as, for instance, the crushing of old slag to extract leftover copper pellets, or the collecting of ores at earlier mining camps (Site 37). In this late period iron smithing was also fairly widespread in the Arabah, mainly in army camps (Sites 11, 46, 213) and road builders' installations (Site 224).

Timna excavations 1964–70

It took several years of investigation of the sites, and detailed study of the archaeological and metallurgical finds, to sort out the chronological sequence of the Timna mining activities. Despite this, reliable, absolute dating of the metal industries in the Arabah was only achieved in 1970, at the conclusion of a series of systematic excavations.

The project of the systematic Timna excavations was conceived in 1963. In the report on the ancient copper industries in the western Arabah, published in the *Palestine Exploration Fund Quarterly*, 1962, the proposed dates and processes of the copper industries were based entirely on surface investigation. Further study of the metallurgical material collected and also the comparison of the pottery from excavations in Israel and neighbouring countries, plus, not least, criticism by fellow archaeologists, made it imperative to try and obtain more information. The aim of future excavations was to find the actual smelting furnaces and workshops, a good collection of stratified flints and pottery, and to obtain indisputable absolute dates for the Timna industries and sites. Excavations started in 1964 at the

Iron Age I smelting camp 2 in Timna. In 1965 Site 39, a Chalcolithic smelting site, was dug; in 1966 a return to smelting camp 2 was made for a second season. Also in 1966 a more detailed exploration of the Arabah was begun, which continued until the Six Day War opened Sinai to archaeological research. After three seasons' work in Sinai excavations were renewed in the Arabah in 1969. In February a Roman smelting camp (Site 28) was excavated at Beer Ora and immediately afterwards Site 200 in Timna, which turned out to be an Egyptian mining temple dedicated to the Egyptian goddess Hathor.

In 1970, just before an additional season in Sinai, the Mameluke site 224 at Ras en Naqb in the mountains west of Elat, was excavated and a unique iron smithy was uncovered.

As the Timna excavations and the Arabah and Sinai explorations are in fact complementary investigations the following chapters are presented against the overall background of each respective period and its major sites in the Arabah and Sinai.

II

Chalcolithic Copper Smelters The Beginnings of Metallurgy in the Arabah

Fig. 4

The earliest traces of human occupation in Timna, in fact the earliest artifacts and ruins found so far in the south-western Arabah, belong to the Chalcolithic period, the fourth millennium BC.

Already in January–March 1959, during the Expedition's first explorations at Timna, a number of flint tools were picked up on the slopes of west Timna, near the copper-ore-bearing Middle White Sandstone formations, and especially near mines 22 and 25 on the northern slopes of Har Timna, and along the foot of 'King Solomon's Pillars', at the south-west corner of Har Timna. Geologists and surveyors of the modern Timna copper mines also collected a number of stray flint tools on the low hills located at the east side of the Timna Valley, going down towards the Arabah, and in the area of Nahal Nimra and Harei Hakhlil. Yet, the realization of the existence of a prehistoric copper industry and the rather extensive settlement in the area was a consequence of the identification and dating of sites 29 and 42, on the south side of Nahal Nehushtan.

Fig. 5

Site 29 – a mining camp

Plate 4

In April 1959 a site in Nahal Nehushtan was re-investigated; it had been previously excavated in 1956 by M. Dotan and identified as a twin burial, including a male with a trepanned skull. It was dated by Dotan to some time between the sixth century BC and the third century AD. The tomb was a rough tumulus built of local red sandstone, with the exception of one piece of granite built into the inner wall of the burial cist. Upon removal, this stone was recognized as part of a larger mortar. Around the tomb, and on the slope below, a large number of flint tools were found. Subsequent investigation in the vicinity of the tumulus revealed the existence of a Chalcolithic mining camp, Site 29 on the survey map, with the actual copper mine, Site 42, nearby. Site 29 (G.R. 14769076) consisted of five groups of large crushing mortars and anvils, located on a slope of approximately 100 × 80 m. above Nahal Nehushtan, and was found covered by a thin layer of wind-borne sand. Numerous stone hammers, flint tools and a quantity of copper ore nodules were found dispersed

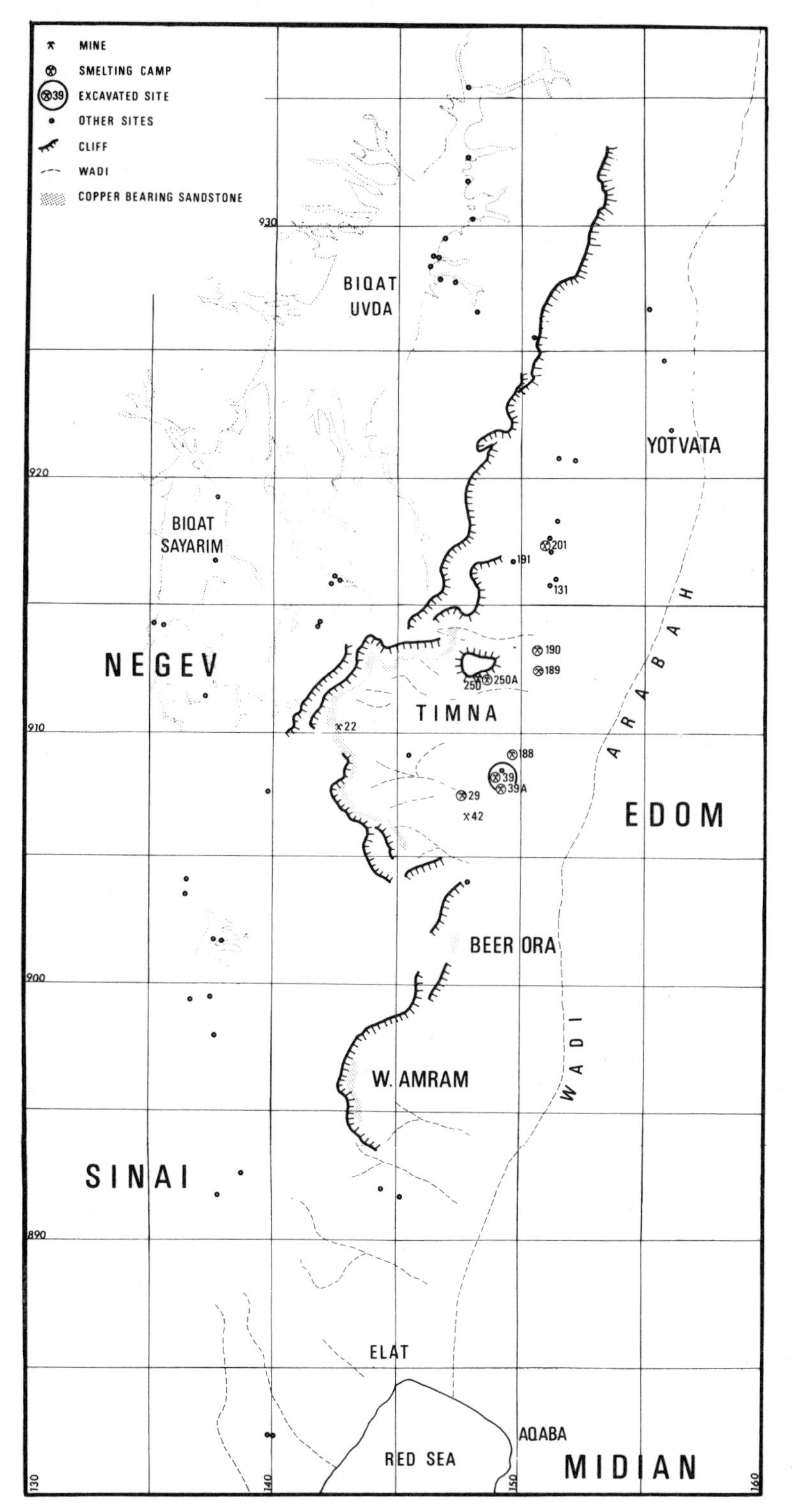

4 Chalcolithic sites in the Arabah and adjoining Negev mountains

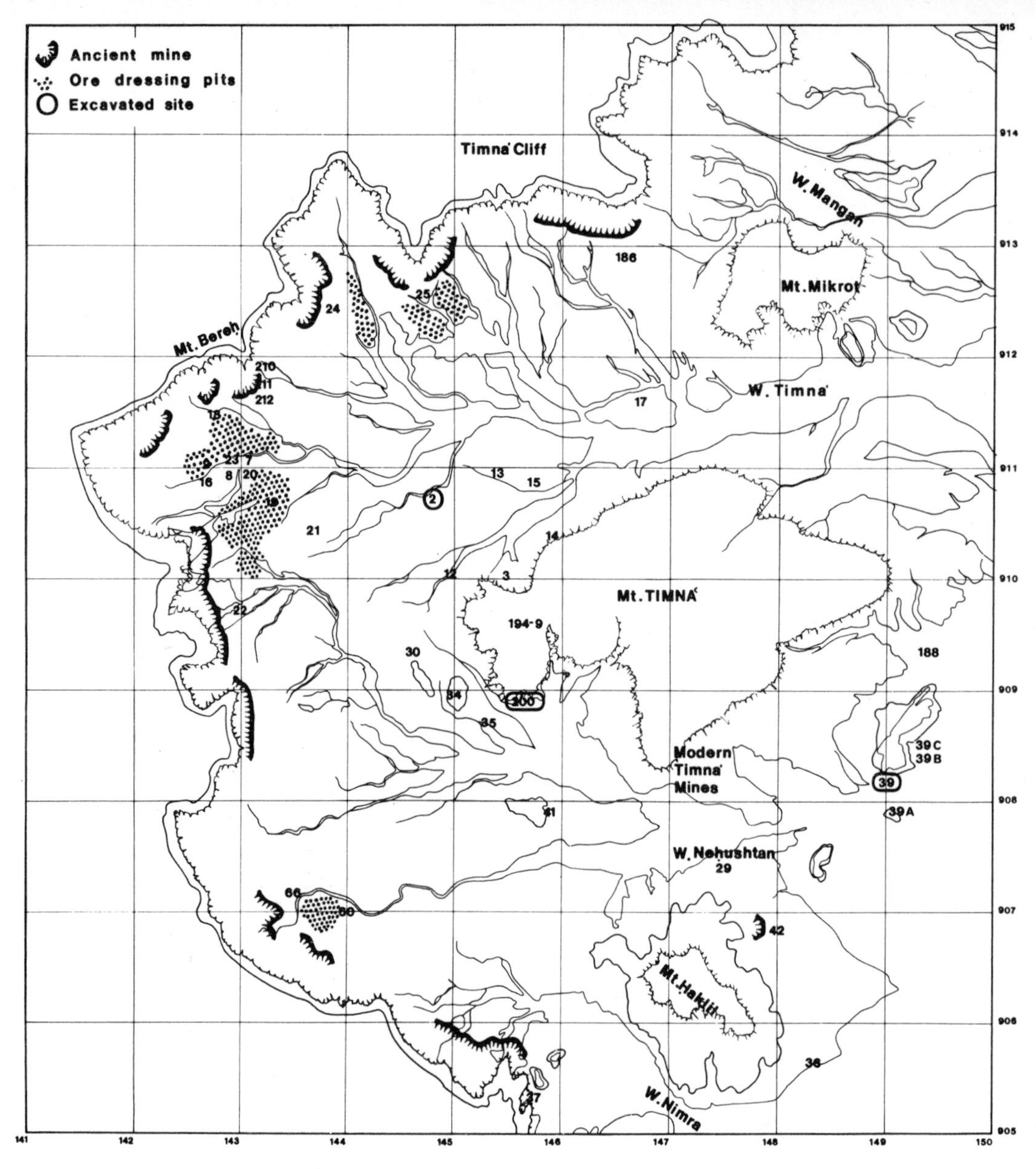

5 *Map of the Timna Valley*

Plates 6, 7

between these large crushing implements. The flint tools could be dated as Chalcolithic, having affinities with material from sites in the Wadi Ghazzeh (Site H and others), Bir Abu Matar and others in the northern Negev. The site was an ore-dressing workshop belonging to the nearby mine 42.

Site 42 – a Chalcolithic copper mine

Plates 2, 3

Site 42 (G.R. 14789067) was found approximately 900 m. south of Site 29, and is a low Middle White outcrop in the hills of Harei

Hakhlil. This greyish-white outcrop contains copper ore nodules, mainly chalcocite with malachite and some chrysocolla, the same as were found at the dressing site 29. The mining wall shows a strange, artificial, round opening about 1 m. deep, surrounded by a deep groove, similar to the 'Schramm' around the opening of the ancient salt-mining sites of Europe, especially at Hallstatt. The opening was too narrow to have served as a mining shaft and we have so far no explanation for it. A heavy (2·2 kg.) mining axe was found near the opening, made of grey basalt. It is approximately 19 cm. long, 12 cm. wide, and grooved all round, to facilitate its attachment to a wooden handle. This is the only mining tool of its kind found in the Arabah, but similar mining tools have been found at many ancient mining sites in Europe, Africa, Sinai and even in North and South America. It seems that this type of tool answers particularly well the technological requirements of early mining and was independently invented in many mining countries.

Plate 3

Fig. 6

Plate 5

The identification of Sites 42 and 29 as a Chalcolithic copper mine and an ore dressing camp gave archaeological meaning to the previous stray finds of flint tools near some of the Middle White formations of Timna. It became clear that the Mesozoic copper ore deposits were already exploited in the fourth millennium. Only the discovery and subsequent excavation of a Chalcolithic smelting installation at Timna would complete the picture of a well organized Chalcolithic copper industry in the Arabah.

Site 39 – a Chalcolithic smelter

Site 39 (G.R. 14939085) was identified as a Chalcolithic smelting site only after repeated investigations and considerable thought, in the field as well as in the laboratory. The reason for these prolonged investigations was mainly the fact that no Chalcolithic copper smelting site had ever been found elsewhere and the basic research had to be carried out before sound technological and chronological conclusions could be reached. As a matter of fact, these were obtained

Plate 8

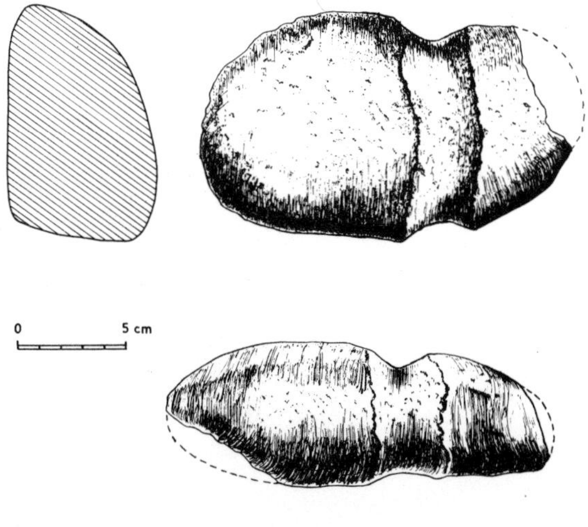

6 Chalcolithic mining axe (Plate 5) found in the Timna copper mines

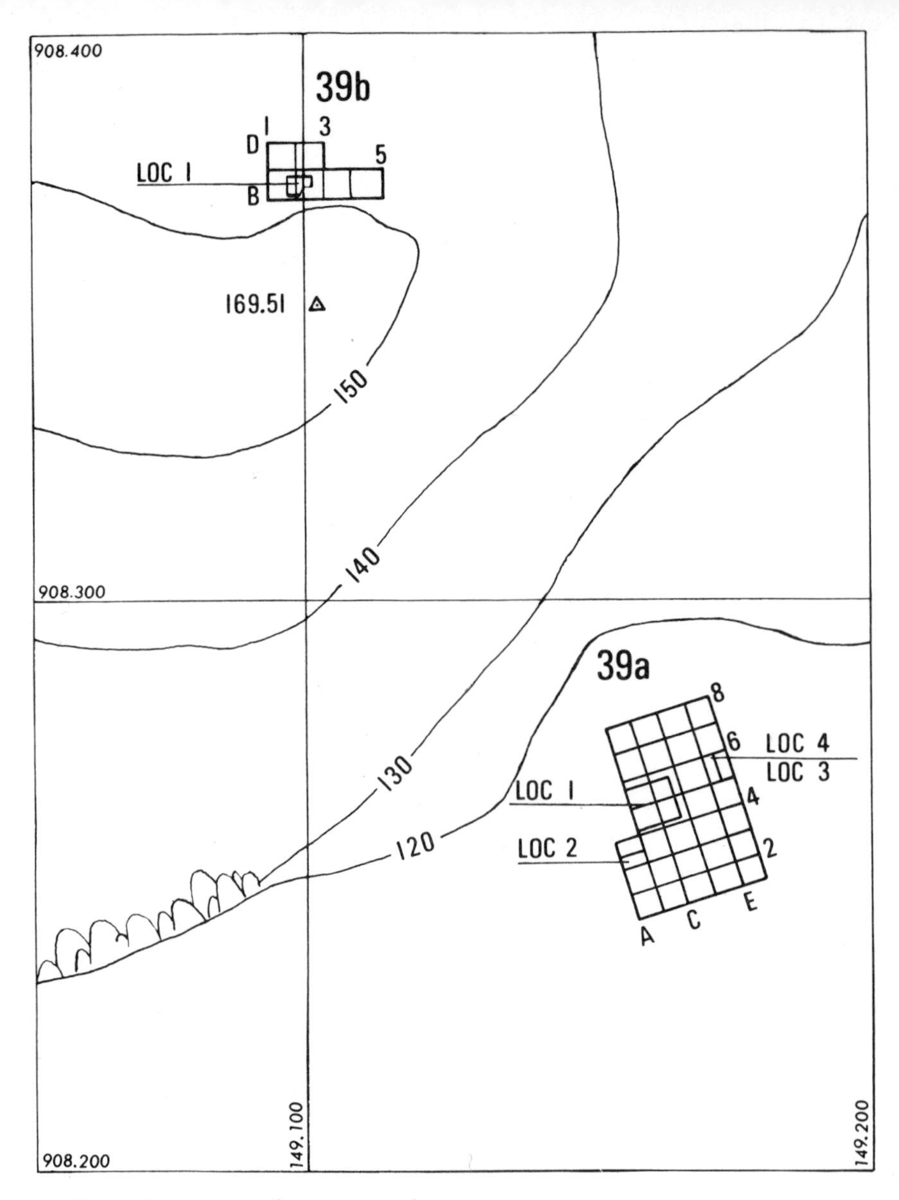

7 *Site 39: Location map of areas excavated*

only by excavating Site 39 in 1965 (five years after the initial investigation).

Fig. 7

The site is located both at the top and at the foot of a solitary hill (approximately 30 m. high), mainly of granite and some red Nubian sandstone, standing on the edge of the Arabah, just north of Nahal Nehushtan. On top of the hill (Site 39b) no structural remains were found but a quantity of very rough, porous, slag pieces, containing many visible copper prills, were indicative of ancient smelting. All the slag pieces had been broken up into small fragments and many such pieces were found at some distance from the actual smelting area. A large number of flint fragments and tools, as well as some rough sherds, were found among the slag.

Structural remains (Site 39a) were located on a slope below at the south-east end of the smelting hill. They consisted of an oval, wall-like assembly of rough field stones, and three tumuli. Here, a number of small flat mortars, many pounding pestles and hammers, flint tools and pottery were collected on the surface. Some copper ore nodules and a few bits of slag were also found. From the tools collected around the structure we could conclude even before excavation that this site served to prepare the actual smelting mixture for the smelters operating on top of the hill above. Flint tools and pottery made a Chalcolithic date for Site 39 most probable.

Plate 8

Plate 10

The excavations at Site 39 in 1965 had a dual purpose:

1 To find technological installations connected with the smelting of copper and thus enable the details of Chalcolithic metallurgy to be reconstructed.
2 To secure enough stratified material to help reliably date the site and the contemporary copper mines and settlements of the Arabah.

Although a quantity of datable flint implements and pottery occurred on the surface it was, nevertheless, considered necessary to provide stratified evidence for the proposed dating of the site to the fourth millennium BC.

Before excavation, the remains of Site 39a appeared as an oval-shaped enclosure, 29 × 23 m., with the width of the wall about 1·40 m., and three flat tumuli of 5, 7 and 8 m. in diameter.

We decided to excavate the largest tumulus (Locus 1), located inside the enclosure, and also dug several small trial trenches, one in the tumulus (Locus 2) south of Locus 1 and two more across the 'enclosure wall' itself (Loci 3, 4).

Fig. 8

Locus 1 was first cleared of debris and wind-blown sand, accumulated loosely on top of what appeared to be a carefully laid cover (4 × 5 m.) of rough, medium-sized field stones. But the stone cover did not extend right across the whole tumulus; it left a circular area uncovered in its centre. Digging down beneath the cover only revealed fine, wind-borne and sterile sand. After approximately 40 cm. a hard sand level was reached which must have been a habitation-surface. Here, right in the centre of the tumulus, was found a curious arrangement of small flat stones (7–20 cm.) forming a round 'floor', about 80 cm. in diameter, surrounded by a thin layer of very fine, greenish, sand. This arrangement resembles a pebble group

Fig. 9

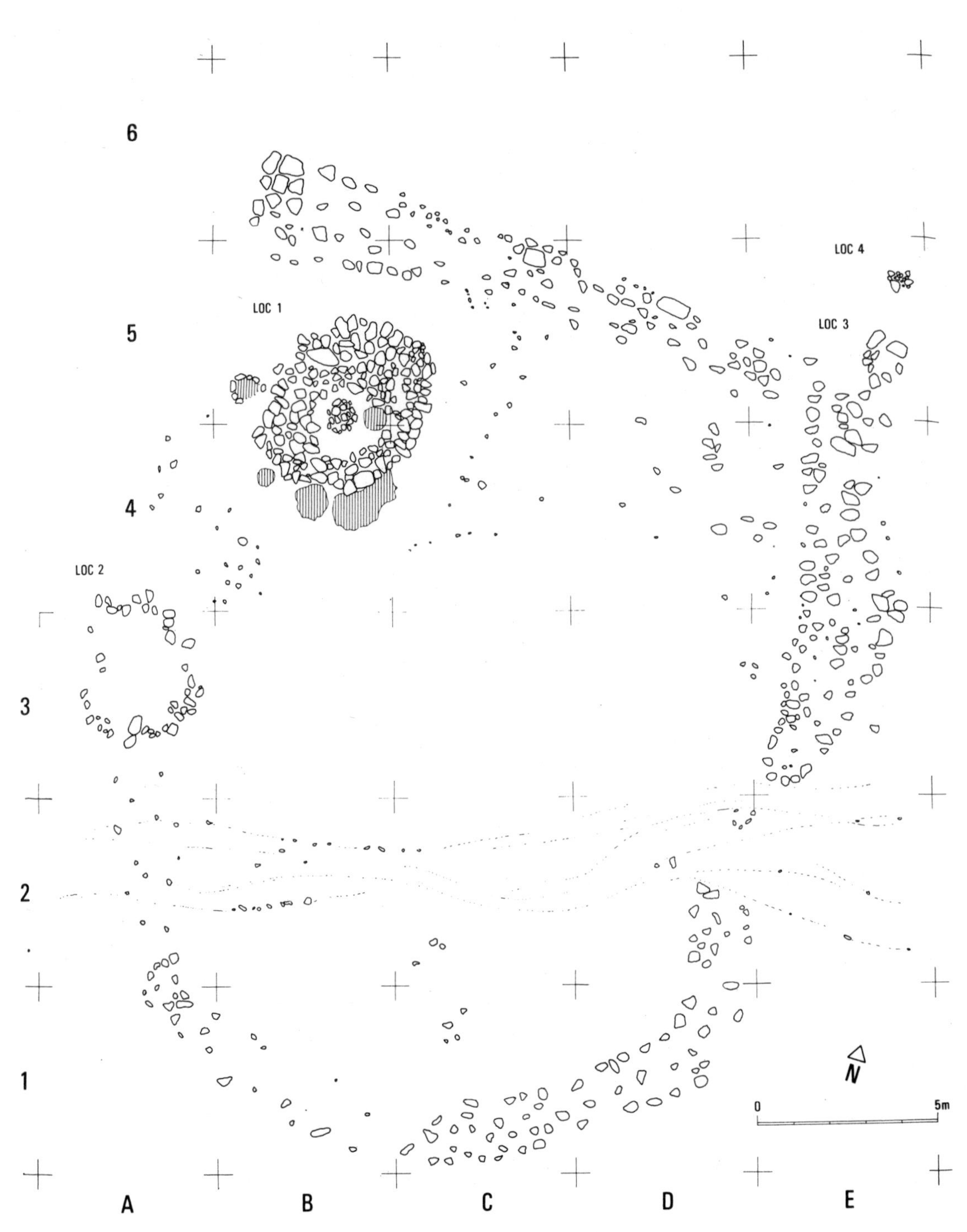

8 Site 39a: A Chalcolithic copper smelter's habitation area, partly excavated

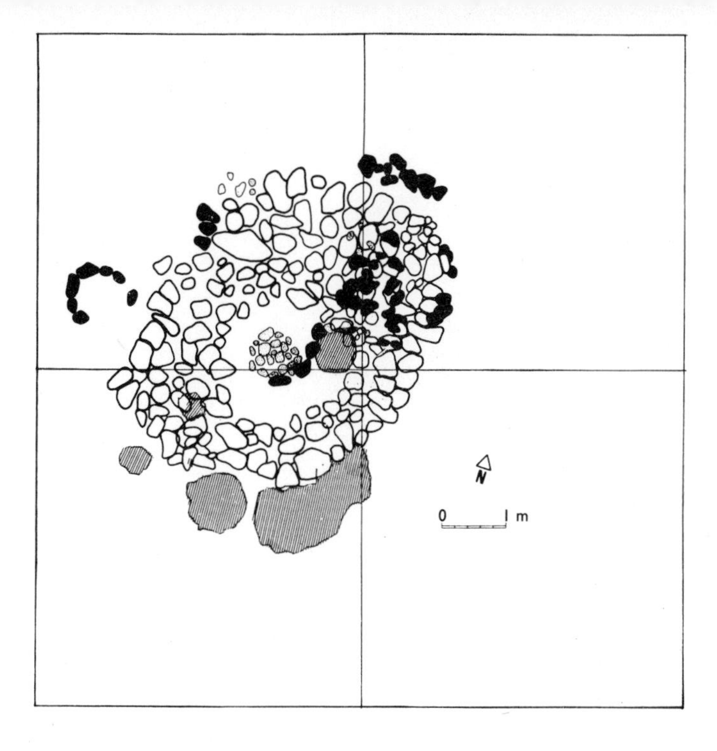

9 Site 39a: Plan of Locus I showing different phases of excavation

found at Tell Abu Matar, near Beer Sheba, but, contrary to the group at Abu Matar, these stones bore no marks of any kind. Next to the group of pebbles was a pit, approximately 25 cm. deep, full of dark black charcoal ash. A small quantity of bones was found nearby.

No other structural remains were found within the tumulus, but the same hard sand surface was found everywhere, both inside and outside the tumulus borders. On this surface were flint implements and flakes as well as some pottery. On the same floor also rested the circular enclosure-wall which appeared after clearing the top stone 'cover' overlying the wind-borne sand layer. An entrance to the enclosure on its west side had a small stone circle, apparently a fireplace, close to it on the outside. The excavation of Locus 1 caused, initially, quite a problem in interpretation, because the stone cover on top appeared to be carefully laid. Yet, as the sand layer underneath the 'cover' was without doubt a result of protracted wind-borne filling-in, it was difficult to understand the connection between the barren centre patch in the 'cover' and the round group of small stones under 40 cm. of wind-borne sand. Only after careful study of sections and measurement of the position and the degree of inclination of the individual stones of the 'cover', was the following conclusion reached. The original structure was a circular enclosure of about 1–1·5 m. in height; in its centre was the small stone arrangement, with a pit next to it. At some time the site was abandoned and wind-borne sand started to settle overall. Subsequently the wall

Fig. 10

Fig. 11

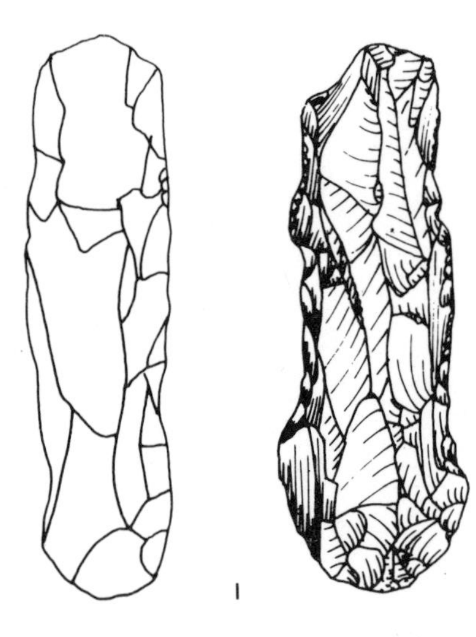

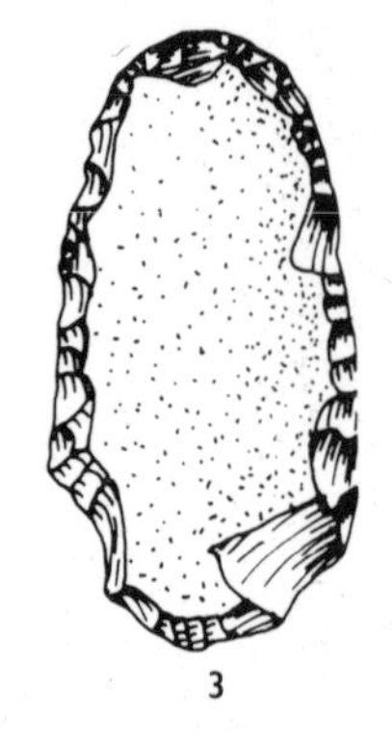

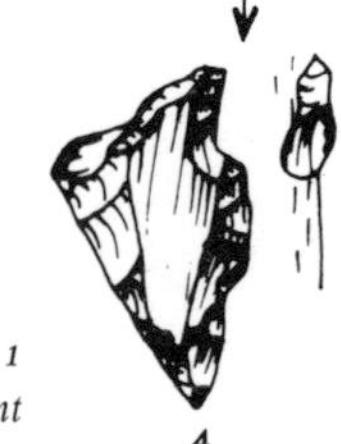

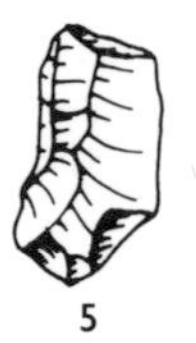

10 Chalcolithic flint implements from Site 39; 1 Pick; 2 Convex side scraper; 3 Elipsoid tabular flint scraper; 4–6 Burins; 7 Hammerstone; 8 Core; 9 Axe; 10 Core

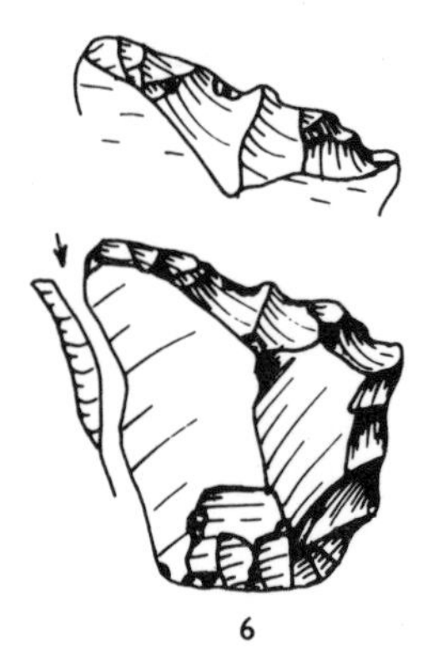

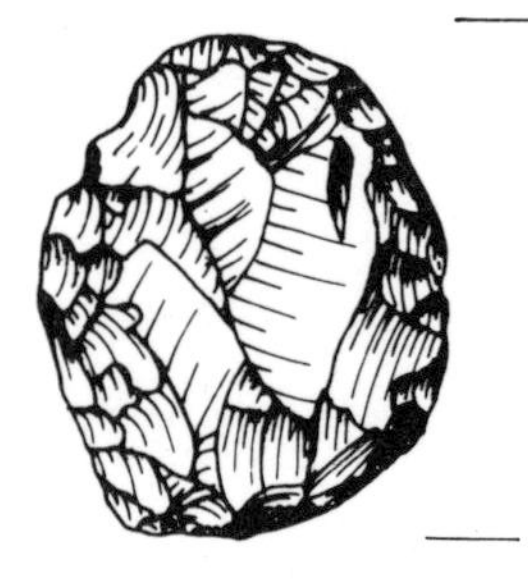

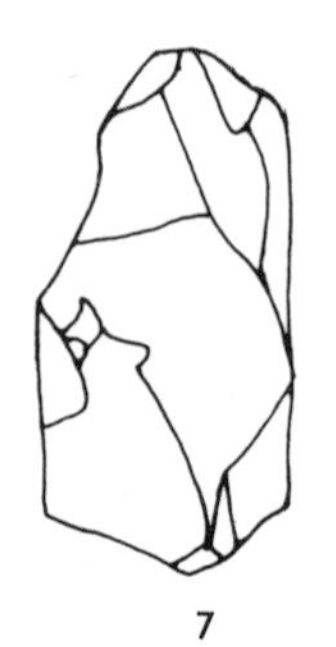

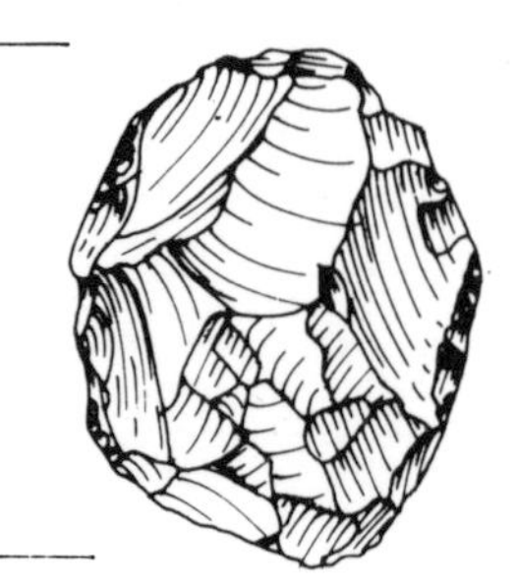

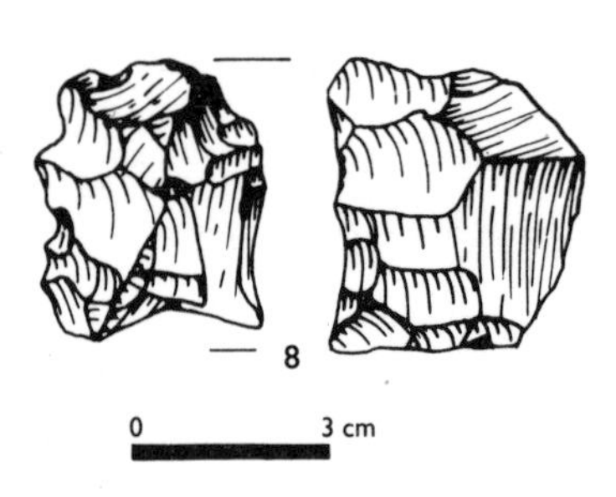

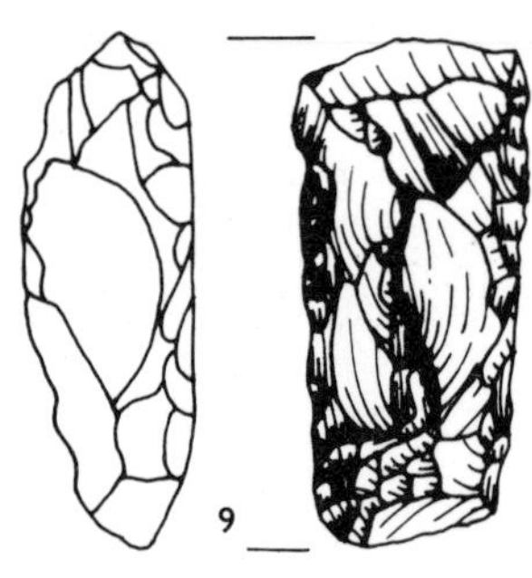

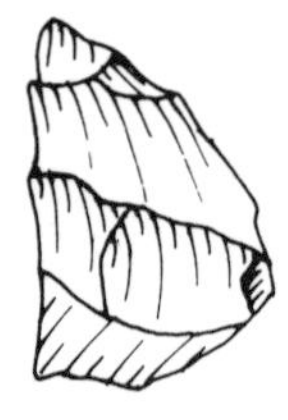

1 Aerial view of the central massif of Timna. In the background is the Wadi Arabah, and in the foreground the Nubian sandstone formation known as 'King Solomon's Pillars'. The location of the Hathor Temple is immediately beneath the 'Pillar' in the centre of the photograph (Plate I)

2, 3 The earliest evidence for Chalcolithic copper mining at Timna was found at the south end of the Timna Valley (Site 2). A small outcrop of white sandstone contained considerable quantities of copper ore nodules. The small circular opening, *right*, too shallow to be a proper mine, is outlined by a curious groove, typical of early mining practice

4 Chalcolithic ore-dressing camp near the Chalcolithic mines, Site 42

5-7 Chalcolithic mining and ore-dressing tools. *Right*, a mining axe found at Site 42 (Plate 3). *Below*, a large granite mortar and a group of mortars as found

8 Chalcolithic smelting Site 39a before excavation. In the background is the Arabah and part of the modern Timna Copper Mines workings. The modern Israeli copper mining enterprise completes the 6000-year-old history of copper working at Timna

9, 10 Remains of a Chalcolithic smelting furnace and, *right*, a crushing tool from the site (39b)

11 A Chalcolithic habitation site in the Arabah to the north of Timna. Many such sites have been discovered and all are chronologically linked by the remains of metallurgical activity, together with finds of flint implements and pottery

12–14 At the head of Nahal Quleb a series of engravings were found cut into the rock-face. They are magic symbols associated with a Chalcolithic shrine. A detail, *above*, shows the central portion of the scene with a group of ostriches on the left, and a curious pair of human figures on the right. The large figure, probably female, is also differentiated by its curious concentric head-dress with two horns

15 Ramesside ore-dressing camp showing saucer-like areas where the ore was gathered and then possibly sifted by winnowing (*cf.* Plate VI)

16 Ancient copper mines at the foot of the Timna Cliffs. The mining walls show traces of shallow digging where ore nodule concentrations occur

← 17 A rock-cut water cistern in the wadi-bed close to the copper mines. Most of the two hundred examples found in the area date to the Ramesside period, and have long since silted up

18 A 17m.-deep cistern near Site 9. The photograph clearly shows the marks of the tools used in its cutting and a series of footholds lead down into its depths. All round the mouth are the marks made by ropes scouring the surface as buckets were raised and lowered. The curious signs nearby have yet to be deciphered

20-22 Tools found in the mining area. The broken granite mining hammer is of the Ramesside → period. The pestle and crude mortar are of a basalt-like stone originating from northern Arabia (the Land of Midian). *Below*, a finished copper spear-butt found on the surface of one of the saucer-like ore-dressing installations near the mines

19 Aerial view of Sites 30 and 34 (*cf.* Plate II). Site 30 is located at the foot of the cliff on the right, surrounded by a defensive wall. Site 34 is located on the plateau of the hill in the background, and similarly defended by walling

23-25 Ramesside crushing and ore-grinding tools made of granite and sandstone, as found in the Timna smelting camps. The fragment of a semi-circular rotary quern, *below right*, is the earliest example known

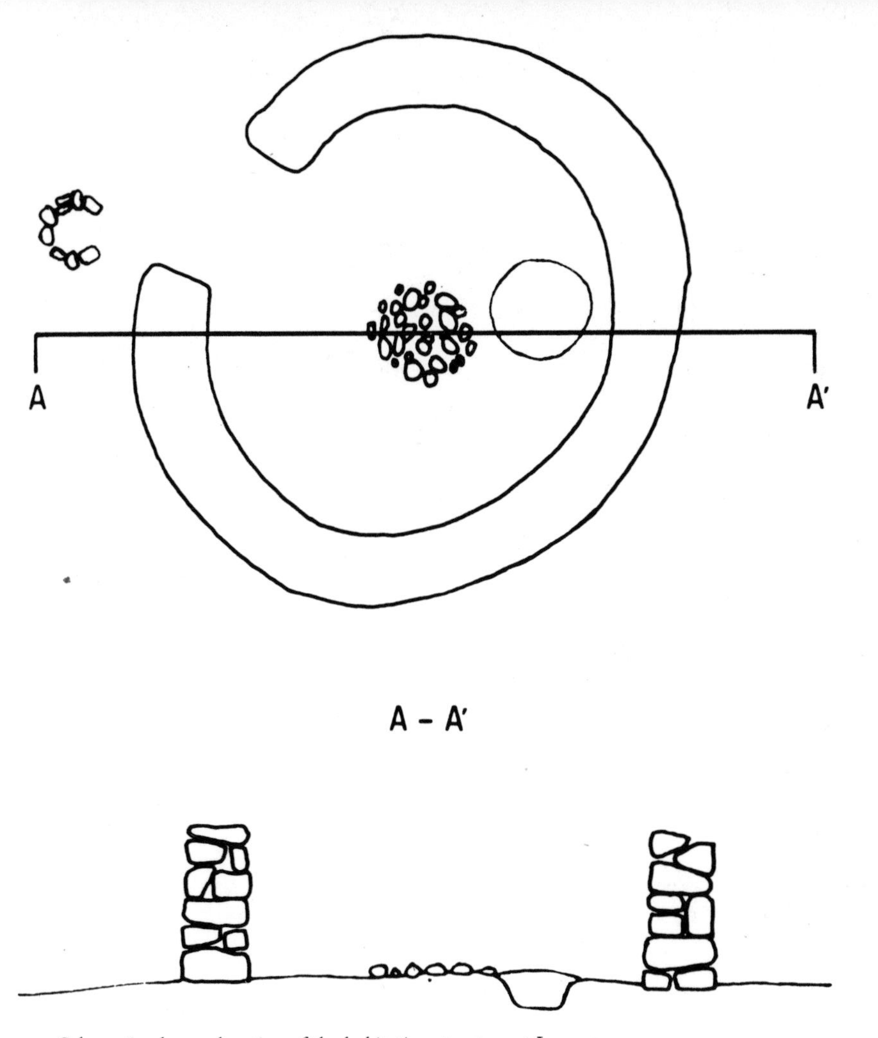

11 Schematic plan and section of the habitation structure at Locus 1

started to give way and some of the stones from its top fell down and were found as debris in the sand-fill, mainly at the north end of the enclosure. Only after about 30–35 cm. of sand had accumulated inside and around the enclosure did the whole structure collapse, perhaps as the result of an earth tremor, and fell on top of the sand-fill, forming a stone 'cover'. The stones falling from the collapsing walls did not reach the centre of the enclosure, which therefore remained bare. This is why the barren patch in the 'cover' fitted accurately the stone arrangement on the floor underneath the sand-fill. Locus 1 seems to have served as a habitation site as no signs of any other use were found.

LOCI 2–4

The trial trenches at Loci 2–4 proved that the large oval-shaped enclosure of Site 39a was not a continuous structure but a line of small workings and fireplaces. Locus 2 may have been a structure similar to Locus 1 but it could not be fully excavated because a stream of sulphuric acid from the modern copper works covered most of it. Loci 3–4 were small fireplaces consisting of some stones and a lot of ashes. Stone tools, hammers and mortars, as well as a quantity of copper ores found before and during the excavation, make the connection of Site 39a with the copper smelting Site 39b quite certain. Thus the conclusion reached from the finds made during the first survey of this site, that it was a working camp for the preparation of the smelting charge for the copper smelters located on top of the hill above, was proved by excavation.

SITE 39B

Fig. 12

The actual site of the copper smelting (39b) was located on top of the hill, only about 100 m. from working camp 39a. Before excavation nothing but small pieces of slag of a peculiar rough kind and some flint implements and sherds indicated a smelting site. The slag was dispersed on even ground behind a narrow ridge, which formed a rocky outcrop on the top of the hill. One particularly large stone, obviously brought here for some special purpose, drew the excavator's attention, and it was decided to excavate an area of 3 × 4 m. around it. At only about 10 cm. below the present surface of red sand there was burnt grey sand mixed with slag fragments and small

12 *Site 39b: Plan and section of the Chalcolithic copper smelting furnace*

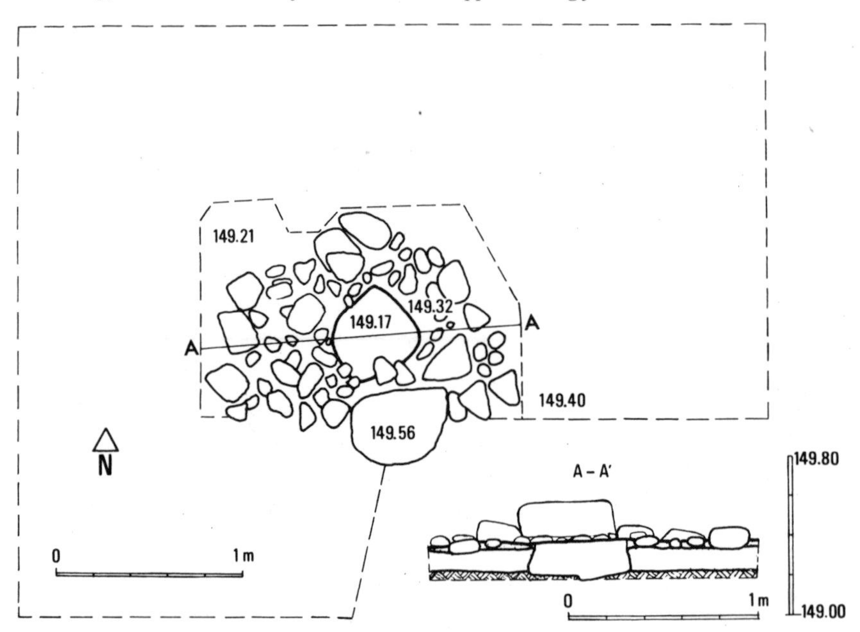

burnt red sandstone pieces. Below this was a pit filled with burnt material containing about 2·5% copper and more small copper smelting slag. At a depth of 25–30 cm. was a burnt, grey bottom consisting of siliceous material and clay. It became evident that here was an actual copper smelting furnace. In fact, only the lower part of the furnace was found; its upper part, which must have been demolished at the time the furnace went out of use, was found as debris scattered around. This furnace was of a bowl type, 45 cm. in diameter, and dug into the hard red sand layer. A circular stone wall of compact sandstone had been built up from the furnace bottom to an estimated height of 50 to 60 cm. Signs of intensive heat were evident at the furnace itself and for about 50 cm. around it. No 'tuyères' or bellows were found, but it seems likely that the large stone, mentioned above, served as a base for the operator of a blow pipe. A working surface with slag, charcoal, flint tools and some pottery *in situ* found at the level of the rim of the bowl, was cleared for some distance around the furnace. Many examples of slag and ores were secured for subsequent metallurgical examination.

Plate 9

The stone-built, copper smelting furnace of Site 39b at Timna is, as yet, the earliest copper smelting installation found. It is very primitive in its construction and performance and may indicate the very beginning of extractive copper metallurgy.

Chalcolithic habitation sites, and more metallurgy, in the Arabah

Despite the fact that more slag concentrations with Chalcolithic flint tools and pottery were located in 1960 on some of the small hills along the edge of the Arabah, both south as well as north of Site 39, Chalcolithic Timna was thought to be only a short, episodic chapter in the history of the area. This view was upheld after the 1965 excavations because at that stage the Chalcolithic remains were still found in apparent isolation; no settlements or habitation sites appeared nearby, or even at a reasonable distance, to have served as a social and economic hinterland for the Chalcolithic copper smelters of Timna. It seemed then that a small group of itinerant metalworkers penetrated into the area of Timna, seemingly from the east, to try and exploit its valuable copper ore deposits. Timna appeared at that stage as an isolated island of copper production surrounded by a huge, empty, desert. However, subsequent detailed survey of the south-western Arabah and the mountains of the southernmost Negev, conducted during 1966–67, changed fundamentally this preliminary conception of an insular Chalcolithic metallurgical Timna. Thirty-eight Chalcolithic habitation sites were discovered in the area explored, resulting in a new understanding of the Chalcolithic settlement in the area. It is now evident that several waves of semi-nomadic, pastoral tribes invaded the Arabah during the fourth millennium BC. No traces of any earlier occupation have yet been found in this southernmost tip of Israel, and the Chalcolithic settlers must be considered the first sedentary occupants of the southern Negev and, as has been proved since 1967, also of adjoining South Sinai.

We can no longer speak of a short-lived, sporadic, exploitation of natural wealth, but must now consider the Chalcolithic–Early Bronze Age occupation of this huge desert, from the Nile Delta to the Arabah, and probably well into the Transjordan Plateau and north-west Arabia, as the earliest and, as the survey proved it to be, the most intensive (and for many parts the only) phase of the history of sedentary occupation in this area. It has already been suggested (in the preliminary publication of the Sinai Survey) that one may see here in this twin-sided, large-scale, movement of copper-smelting and turquoise-mining people into Sinai, the natural meeting place of African and Asian culture in Predynastic times, a cultural land bridge between Africa and Asia. A few Pre-pottery Neolithic B sites found in Central and South Sinai indicate even earlier contacts between Asia and Egypt.

Returning to Timna, this valley of copper may be considered as just one of many Chalcolithic copper industries in the great desert, but perhaps one of the largest and richest copper deposits and therefore a most important centre of Chalcolithic copper production. The Chalcolithic settlement pattern in the Arabah and Sinai is clearly that of pastoral semi-nomads having not only a good knowledge of copper metallurgy, but also of ceramics, stone-tool making and, where arable land was available, of small-scale farming. Most of the Chalcolithic settlements in the Arabah, located from Beer Ora to Kibbutz Gruphit, are small groups of circular stone enclosures (3–12 m. in diameter), some were used as sheep and/or goat corrals, others perhaps foundations for huts or tents. Sometimes there are large, village-like clusters of structures, but the general pattern is similar to present-day Beduin camps, groups of one or two tents belonging to one family; sometimes several family-units are grouped together and form a common camping area.

While almost all Chalcolithic sites in the Timna area are either mining or smelting sites, forming almost a separate region of copper mining and smelting, the vicinity of Yotvata, with its springs and swamps (called Malehat Yotvata), shows a concentration of habitation sites. A second area of Chalcolithic habitation and food production is located in the mountains north-west of Timna and Yotvata. Here, clusters of Chalcolithic enclosures are found along the edges of the extensive loess flats of Biq'at Uvda and Biq'at Sayarim, which are very fertile in years of rain, and suitable for dry farming. Although we have evidence for Chalcolithic shallow well-digging, mainly in the vicinity of Beer Milhan and Beer Metek (area of G.R. 916143), with large Chalcolithic villages nearby, it appears that the farming sites around the flats were occupied only during the rainy season, because there is not sufficient drinking water available for permanent occupation. Perhaps we may see here the earliest evidence for a seasonal interchange but parallel occupation of two different landscapes by the same people. The food-growing members of the community moving up to the farming areas in the mountains for the rainy season, only to come down again to the central tribal settlements in the well-watered Yotvata area when drinking water gave out in the mountains.

The Chalcolithic settlement pattern in the Arabah and South Sinai furnishes perfect evidence for the existence of a well organized, semi-nomadic, community, based on specialization of work, from hunting and food-growing to copper working and, consequently, a high degree of technological progress.

'Desert Kites' and a Chalcolithic shrine in the Arabah

Whilst the simple circular stone enclosures of the average Chalcolithic habitation sites show only minor variations, two sites of particular interest should be mentioned. In the alluvial fan of Nahal Quleb, west of the well-watered Malehat Yotvata, were found two stone structures of a peculiar design (Site 131, G.R. 15129159), previously known only on the Transjordan Plateau and in the Arabian Desert, and called 'Desert Kites'. A Desert Kite is a system of two long walls, meeting at an angle of approximately 30°, and forming, thereby, a triangle consisting of two long sides, in this case over 200 m. long, with an open base. At the meeting point of the walls a tower-like, round structure is found. The two Arabah Desert Kites are built close together in order to include between their long open arms a wide stretch of land (approximately 300 m.). The character of these structures and their location make it clear that they were built as traps for gazelles and, perhaps, ostriches. The Desert Kites in Transjordan have so far not been dated, but Chalcolithic pottery and flint implements found in close proximity to the stone walls, and also in the corner tower, of the Arabah Desert Kites

Fig. 13

indicate a date within that period. This dating is underlined by several more Chalcolithic sites nearby, one of which, Site 191, appears to be a Chalcolithic shrine.

Site 191 (G.R. 14979166) is a rock formation of Middle White Sandstone, located at the head of Nahal Quleb, west of the Desert Kites and is conspicuous from afar because of its grey-white colour. The rock formation is approximately 10 m. high, cut by small gullies into several thick, pillar-like blocks of smooth, roundish appearance. A low stone enclosure was built against the rock face, today ruined and much disturbed by Beduin camping. Each end of the semicircular enclosure wall touches one of the two central 'pillars' of the rock formation, with a narrow, man-wide, canyon in between the two 'pillars'. A relatively large quantity of Chalcolithic sherds and some flint implements were found within the area of the enclosure. Inside

Plate 12

13 Map of Nahal Quleb with 'Desert Kites', a hunting trap for gazelles and ostriches

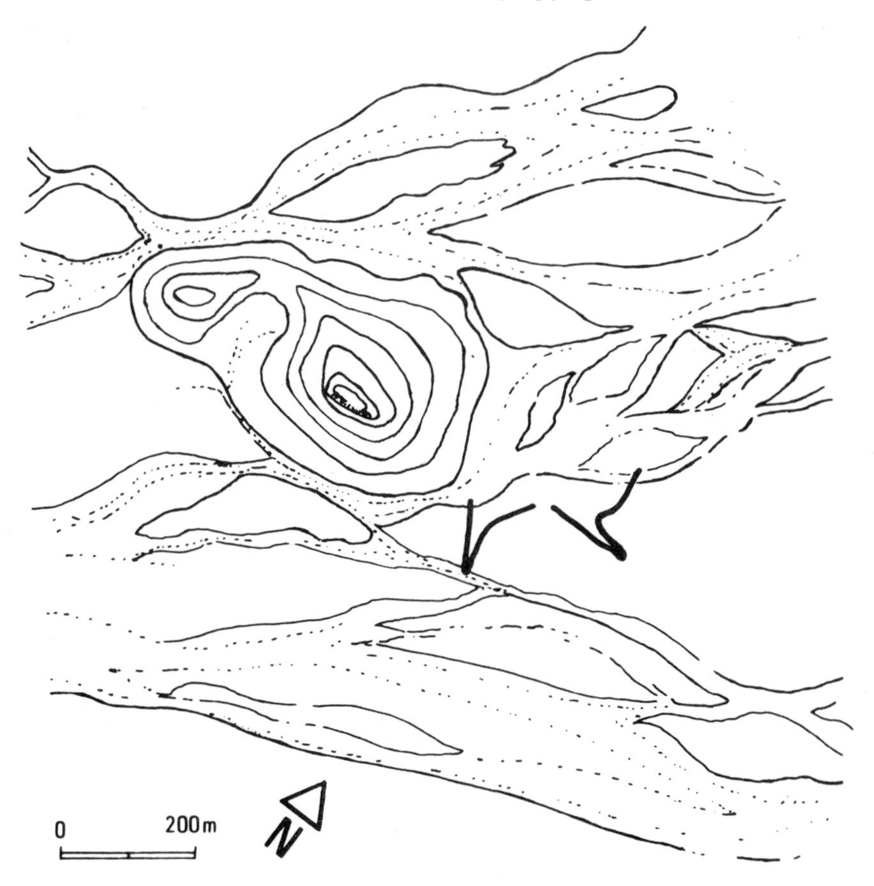

14 Rock-engravings at Chalcolithic Site 191

the narrow canyon there is a group of unique engravings on the rock wall. Nothing like them is known in the Arabah, Sinai, or the Central Negev, where many hundreds of rock engravings have been found and recorded during the last fifteen years. Deep inside the canyon, one group of engravings represents a strange pair of 'human' figures, perhaps the divinities worshipped here. The right figure, 55 cm. high, is reminiscent of the geometrical character of the Chalcolithic frescoes at Ghassul. Possibly female, it has concentric circles on top of its head, drawn as three short incisions, and is crowned by two horns. The body is very schematically drawn, with one of the hands showing three fingers only. The left figure, 45 cm. high, perhaps a male, has the same style head with three short, deep lines but no further emblems. Its body is drawn like its female companion, but it has only one arm and a straight line indicates its two closed legs. Double lines, horizontally attached to the body below the shoulders, are perhaps intended to represent wings. The strangeness of these two figures is most striking, and their location, deep inside the rock-fissure, seems indicative of the importance attached to them by their artists.

Fig. 14

Plate 13

Plate 14

A second, unique, group of engravings is located on the same wall but near the entrance to the canyon and is visible from outside. In fact, it is the first thing noticed when approaching the site. Here, four engravings, 100 cm. high, show ostriches in a schematic, almost abstract, technique, the exaggerated, upright, long necks with a mere indication of the head. The big birds are represented from three different sides and the abstraction of the primitive artist reaches a sophsticated level of abstract realism, giving the very essence of the

I 'King Solomon's Pillars'. The Hathor Temple was found under the rock overhang to the left of the jeep.

II A Ramesside smelting camp (Site 30) seen from the hill above (*cf.* Pl. 19). It is surrounded by a defensive wall with two gate towers. The dark areas within the wall are smelting slag on the site of the copper furnaces. The rough stone structures on the right are the remains of workshops and habitations. In the background are the Timna Cliffs with the ancient copper mines at their foot.

III An early Iron Age copper mine and smelting site in the Wadi Amram. The copper ore is found in the Middle White sandstone horizon, seen as a sloping strata in the cliffs, which reach a height of over 300 ft. Smelting operations took place in the valley below the mines.

IV The 'Amram Pillars' in the Wadi Amram surrounded by white cupriferous sandstone, the site of copper mining in the Roman period.

V The Timna Cliffs with the ancient copper mines in the foreground.

VI Part of a working camp on the slopes below the Timna mines. The saucer-like working areas in the middle distance served as dressing stations for the ore. They are typical of the Ramesside period.

I

II

III

IV

V

VI

VII

VIII

IX

VII A Ramesside smelting furnace in camp No. 2 (*cf.* Plates 35–37).

VIII Copper ore nodules composed of azurite, malachite and chalcocite.

IX High grade fragments of fossilized tree trunks containing iron oxides used as flux in the smelting processes.

bird's form and movement, and no more. The ostrich on the right is drawn front view, its legs as one strong line; the bird to its left is drawn in profile, with the body beautifully carved out in relief and the two legs given as two short lines with almost human feet. The third ostrich, drawn from rear view, has an unusually long neck; its wings are strongly drawn upwards and its two legs represented by two parallel strokes. The fourth bird is much smaller than the others and details of its body are not quite clear. The four ostriches are not drawn on the same plane but more as if a flock of dispersed birds was caught unawares by the artist. A date within the Chalcolithic period is suggested for these two groups of engravings.

All over this wall, next to the two large groups described above, numerous tribal marks (*wasm*) were carved by Beduin, apparently over a considerable period of time, but great care was taken by the latecomers not to harm the early engravings. Some of the Beduin engravings seem ancient, others fairly recent, but all show the same characteristic geometrical style. Several small human figures, with outspread hands and fingers, and a schematic 'tree-of-life' also appear on this wall. Many more small Beduin engravings are carved on the rocks outside the enclosed area, but none are found anywhere else in the vicinity. It appears that this extraordinary, isolated spot, with its Chalcolithic enclosure and engravings (which may be assumed to be a Chalcolithic cult centre or shrine), also aroused the curiosity and attention of the desert nomads who carried on the ancient tradition of rock engravings here, perhaps with a similar magic intent.

III

Ramesside Copper Mines at Timna

The mines
Fig. 15

The Middle White horizon in the carboniferous Nubian sandstone of the Timna Valley is a discontinuous sedimentary formation from 10–30 m. high, interrupted by red Nubian sandstone intrusions and the narrow and rugged side arms of the four main wadis of the Valley. Most of this horizon contains some copper ore nodules but the ancient miners must have noticed that some of the rocks contain more nodules than others and concentrated their efforts accordingly. This is why eight large centres of mining activities were found along the 10 km. long white sandstone formation, operating at the end of the Late Bronze Age and well into the Early Iron Age I, from the fourteenth to the twelfth centuries BC. As the Arabah Expedition's excavations proved, this was a Pharaonic Egyptian enterprise of the Ramesside Dynasties, working in conjunction with the local Midianite and Amalekite tribes.

Plate 16

The mining faces of this period show clear evidence of open cast mining and, sometimes, shallow digging following rich concentrations of copper ore nodules. Many hard flint, granite and gabbro hammer stones and anvils, mortars and pestles, as well as red, saddle-backed, hard gritty sandstone grinders were found at the mining face. Some pottery was also collected there.

Plates 20, 21

Numerous horizontal surfaces below the actual mines were found covered with saucer-shaped hollows, 1·5–2 m. in diameter, called 'plates', and filled with white sand, near which are remains of crude stone structures or shelters. Many grinding tools were found around these plates. These are ore dressing areas for grinding the ore at the mine in order to remove superfluous gangue before smelting.

Plates 15, VI

At all the Ramesside mining areas tubular cisterns were found, up to 17 m. deep and 70–90 cm. in diameter. These cisterns, carved with metal chisels into the sandstone wadi-bed or into the mining face were used for the storage of run-off rain water. Their mouths showed rope marks and footholds that had been cut into the wall to facilitate cleaning out the accumulated silt. A large quantity of water must have been stored in the 200 or so cisterns found in the Timna Valley.

Plate 17

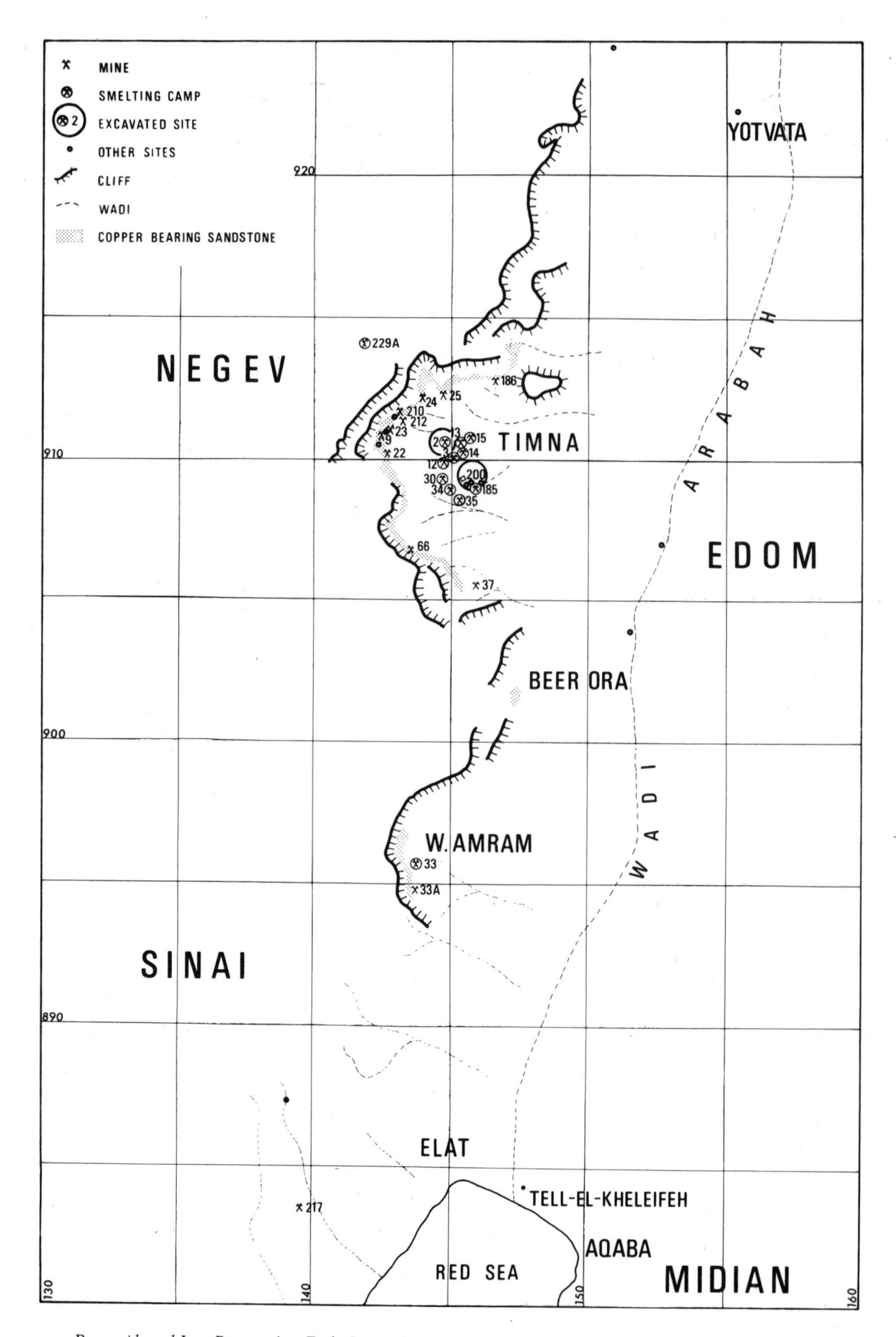

15 *Ramesside and Late Bronze Age–Early Iron Age sites in the Arabah and adjoining Negev mountains*

One of the cisterns at Site 9 was found split by a fissure, apparently caused by an earthquake. It was empty of silt, 17 m. deep, and its mouth shows deep rope marks. Near the mouth a sign was carefully cut into the rock and behind the cistern was found a line of much eroded signs or letters, as yet undeciphered.

Plate 18

The smelting sites

Ten camps of the Ramesside period, directly connected with copper smelting, were found in the Timna Valley, set up in a semi-circle west of the Timna massif. As only Site 2 was excavated it is impossible at present to establish detailed chronological data for these camps, apart from the fact that the pottery found on all of them indicates a Ramesside date as the general period for their operation. It would seem certain, however, that not all of the ten camps were in operation at the same time.

Camps 13, 15, 14, 3, 12 and 35 are groups of stone buildings, some consisting of one or two rooms only, others containing several rooms. The house walls stand today up to 1·5 m. high; built of dry stone walling with rough field stones, they rarely show signs of primitive tooling.

Only very small quantities of slag were found in these camps, or on the slopes around, and it is obvious that no copper smelting took place here, but there could have been some casting activities. The stone tools, pottery and installations found indicate that, besides being simple quarters for the smelters working in the neighbouring smelting camps, many of these stone houses must have served as workshops and storehouses. In fact, stores of ready smelting charge, charcoal and decayed foodstuffs were found inside some of the buildings.

Plates 23–25

Actual smelting took place at sites 30, 34, 185 and 2. Here, considerable quantities of copper smelting slag were found, usually thrown on to a heap next to the actual smelting-installations which, however, are today not visible except in the excavated camp 2.

Camp 30, located in Nahal Nehushtan, at G.R. 14479093, next to a solitary high mountain, is surrounded by a massive semi-circular stone wall, and is approximately 80 m. in diameter. This wall abutted against the sides of the mountain and even climbs up its sides for several metres. The wall is 1·5–2 m. high and 1·2 m. wide. On its north side, two collapsed towers guard the only entrance to the camp. A large heap of big slag pieces of semi-circular shape is piled up in the western half of the camp and here, undoubtedly, smelting took place. The eastern half of the camp shows several

Plates 19, II

destroyed workshops. Very many saddle-backed red sandstone querns and granite and flint hammerstones were collected and also a quantity of charcoal. A low stone dam built across a small side arm of Nahal Nehushtan diverted the run-off floodwater into a rock-bound cistern, cut into the hillside about 40 m. west of the smelting camp. Two rock-carved troughs, next to the cistern, must have been used for watering the pack animals.

In 1969 a small trial hole was excavated next to the large slag heap. No structure was found but several very large clay protectors for bellows-ends came to light, the like of which we had not found anywhere else in Timna. Similar tuyère-ends were found by the Sinai expedition in 1969 at the large copper-smelting camp near Bir Nasib and identified as early New Kingdom smelting remains. It therefore seems likely that camp 30 belongs to the earliest phase of copper-smelting in Timna, although future excavation will have to verify this assumption. The fact that camp 30 is defended by a strong circumvallation may point to the same conclusion, *i.e.* at the very beginning of Egyptian exploitation of Timna's copper ore the local tribes were apparently unfriendly and proper defence measures had to be taken. As at Serabit el-Khadem in Sinai the local tribes, first considered as the enemy, later on became partners or employees and defence walls were no longer necessary.

Similar considerations must be applied to camp 34 (G.R. 14509090), located on top of a flat mountain, about 200 m. south-east of camp 30. Here too, a defence wall of rough fieldstones was piled up along the edge of the plateau, wherever the slope could easily be ascended. A tower-defended gate can be found on the north-west side of the hill. The central part of the flat hill top is covered almost completely with broken slag, presumably broken and dispersed by later people who did not themselves smelt copper but only came to extract residual copper pellets from the old slag heaps. At this stage of the investigations sites 30 and 34 are assumed to belong to the initial phase of the Egyptian copper works at Timna, although these major smelting camps could have been in operation also during the subsequent phases. The pottery found on the surface of sites 30 and 34 does not, however, seem to differ essentially from the pottery found at the other Timna sites. We shall return below to the problem of the Timna pottery and its chronology (pp. 152–163), but at this point it is sufficient to note the variance.

Plate 19

The two unwalled smelting camps in the Timna Valley, sites 2 and 185, certainly belong to the later and last phase of Egyptian copper smelting in Timna, but might have been operating earlier.

Camp 185 is a long line of small ruined structures, standing along a narrow side arm of Nahal Nehushtan (at G.R. 14599090), which runs down from the southern slopes of Har Timna. Most of the buildings consist of one or two rooms and a lot of pottery was found all over the site as well as grinding stones, hammers and anvils. The metallurgical activities were restricted to the south end of the camp. Here a quantity of slag indicates copper smelting as well as casting. Camp 185 was not a major smelting-plant but served mainly as a working and habitation site. The pottery found here places the site well within the Ramesside period.

Explorations and soundings 1959–63

Plate 27

Site 2 was first discovered in 1959. Unlike most of the other sites in Timna, it had not been previously located and ransacked by archaeologists and visitors and a large number of working tools, pottery and even some copper implements were found on the surface at the time of the Expedition's first visits to the site. Several big slag-heaps and a lot of slag dispersed all over the very large camp area made it clear that this was one of the major copper producing sites in the Timna Valley. For that reason and because the surface finds, including the pottery, seem adequately to represent all other sites of the same period in the Timna Valley, the main efforts during the following years were concentrated on Site 2; first by minute surface explorations and small soundings, followed by analytical research in the laboratory, and later by systematic excavations.

Fig. 16

Site 2 is located in a southern, small side arm of Nahal Timna, at G.R. 14489107. Coming from the main Timna road, which runs along the northern side of Har Timna, one drives up a low ridge between two dark mountains and faces, rather unexpectedly, a large, beautiful stone 'mushroom' and behind it a small, sandy valley. The mushroom, about 6 m. high, is the erosional product of the strong north winds, prevalent in this area. The valley is formed by three flat slopes, ending in a narrow, red, sandy wadi bed. On both sides of this small wadi, slag heaps indicate the site of smelting. Further concentrations of slag, charcoal and burnt ground, indicating metallurgical activities, can be found at the extreme east side of the valley, where the slope gradually rises towards the mountains. All three slopes, measuring approximately 150 × 180 m., are covered by piles of medium-sized rough stones with many, although completely destroyed, structures still discernable. Some of the rough building stones seem to have been in secondary use for the erection of simple, mostly circular, enclosures. In fact, at the time of the first survey it was not suspected that beneath the ruined primitive structures, there existed large, well-built workshops, stores, and numerous metallurgical installations, in an excellent state of preservation. The structures today visible on the surface are either a late phase of the use of the smelting camp or simply the collapsed tops of previously free standing walls, their lower courses still partly preserved underground.

Two structures in areas A and F, which are an exception to this, were found standing isolated from the actual smelting site; they will be dealt with below, in Chapter IV.

Numerous stone tools – round, square and oblong flint, granite and sandstone hammer-stones, many large mortars and especially

16 Site 2: Location map showing areas excavated in 1964, 1966, and large slag heaps

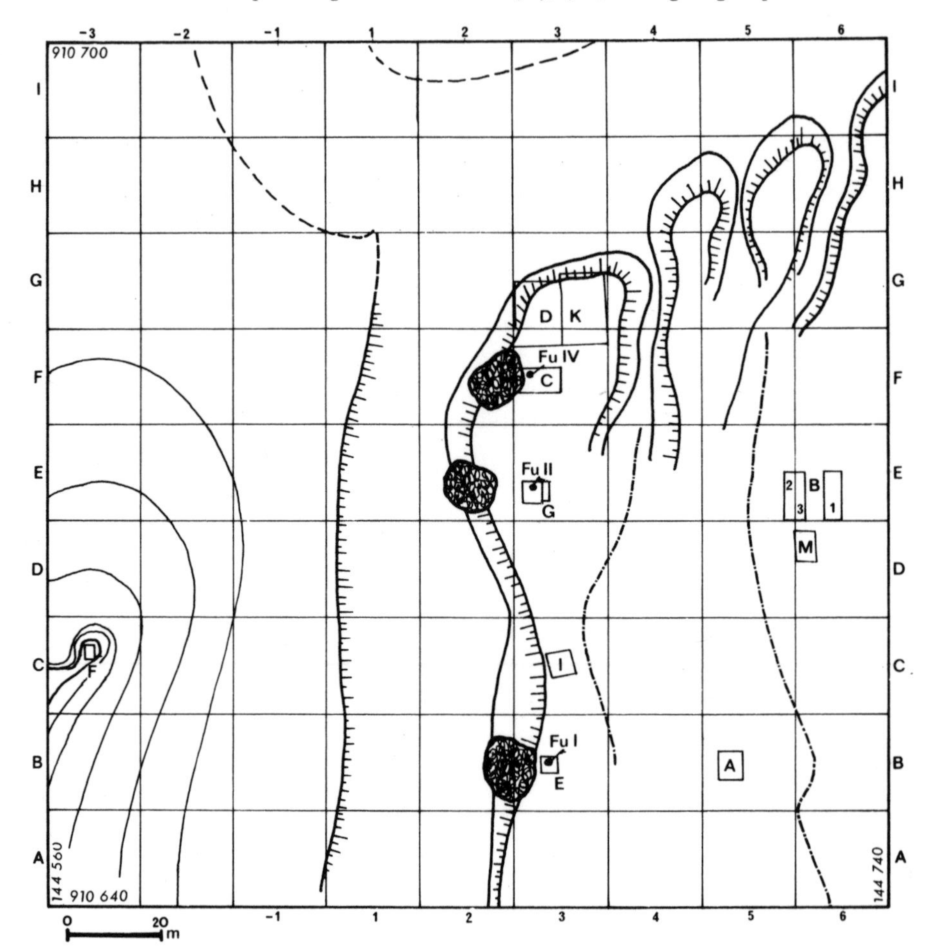

the typical saddle-backed red standstone querns – were found in and around the structures and in the debris. Surface copper finds were, of course, rare but, nevertheless, a complete knife, a spear-butt and many lumps of raw copper were found, besides a quantity of copper pellets which had been extracted from the slag by breaking some of it into smaller pieces.

Plates 23–25

Plate 69; *Fig. 29, 1, 8*

Slag and charred ground on the slope west of the small wadi bed, somewhat outside the main camp site, was proof of metallurgical activities. However, the main location of copper smelting at Site 2 was across the wadi, also along the edge of the slope. Here, three large slag heaps, mixed with clay furnace lining, charcoal, stray ore pieces and clay tuyère ends were obvious signs of copper smelting and, already in 1962, remains of several furnaces were found next to the slag heaps. Small soundings were made in 1963 in areas E and G which helped in the preliminary reconstruction of the smelting process, later fully confirmed and augmented by systematic excavations.

Even at that stage of the investigations it became absolutely clear that copper ore was reduced here to metallic copper in small, earth-bound and partly stone-built, bowl furnaces. The finding in one of the houses of a store of ready smelting charge, in the form of a dark powder, and of the contents of a furnace that seemed to have gone out, showing the actual smelting process – slag on top, and underneath it partly reduced ore and burned and unburned charcoal, mixed with a large number of metallic copper pellets, evidently stopped on their way to the furnace bottom – helped a great deal towards the understanding of the method of copper smelting used in Timna.

A great deal of time was spent on metallurgical research as the generally accepted theories about mining and smelting activities in the Arabah suggested some kind of preliminary roasting of the ore at Timna and the actual smelting/refining was said to have taken place only at Tell el Kheleife, on the shores of the Red Sea. The facts discovered by the survey did not substantiate these theories. First, and most important of all, solid metallic copper was made in the small smelting furnaces of Timna. The ore was crushed by means of the numerous stone-crushing and grinding tools of all sizes (found on the surface), mixed with fluxes, also found at the site, and was then reduced to metallic copper on a charcoal pile inside small bowl furnaces. A bun-shaped copper ingot of about 7 kg., an earlier find in the Arabah, fitted well into this picture of the smelting method and represents the end product of the Arabah smelters.

17 Bun-shaped copper ingot found in the Arabah (wt. 42.2 gm.)

This first reconstruction of the Timna smelting process still did not explain all the facts found at Site 2 but it could, nevertheless, be accepted as a preliminary working hypothesis. Much of the slag in the big heaps consisted of large, circular, lumps of slag weighing up to 40 kg., and most had holes in the centre. All of these slag circles seem to have been broken into several parts at the time they were thrown onto the slag heap. Several of them could be completely reconstructed. A number of questions immediately arose: where and how were these circular lumps of slag formed, inside or outside the furnace, and why was there a hole in their centre? Furthermore, besides the large, heavy and dark coloured slag, many smaller pieces of porous appearance and of light brown-grey colour were found at Site 2. There was no archaeological or metallurgical explanation for the appearance of these two different kinds of slag and it became obvious that these questions could only be answered by systematic excavations of actual smelting installations.

Plate 31

The pottery from Site 2, right from the beginning of the survey, created a complex problem and, although the excavations at the site in 1964–66 produced a lot of stratified material, the absolute dating of the Timna pottery remained for years the subject of protracted arguments and deliberations, to be finally resolved only in 1969 by the excavation of the Timna Temple, Site 200, and the consequent dating of the pottery by the inscriptions found.

All over Site 2, as at most sites in the Timna Valley, sherds of three distinct kinds were collected:

Plates 42, 43

1 'Normal' ordinary wheel-made pottery, plain but well-fired, consisting mainly of many-handled storage jars, carinated bowls, jugs and juglets.

Plates 44, 45

2 Coarse, hand-made, deep as well as shallow bowls used for cooking and domestic purposes, akin to that previously found in the Central Negev Mountains, and named 'Negev-type ware'. Many of the flat bottoms of these bowls show mat-impressions.

Plates 48–54

3 Unique, pink-buff ware, decorated with bichrome geometrical designs (red-brown and black), made of well-levigated, evenly fired clay. Most of these sherds found at Site 2 belonged to large deep and shallow little bowls, with flat bottoms and straight sides, and having an occasional knob-handle projecting from the rim. There were also fragments of deep cups, decorated with bichrome 'Union-Jacks' and similar geometrical decorations, and shallow bowls with a floral design in the centre. No pottery of this kind has ever been found in Palestine but it has been picked up on the

surface of sites in Jordan, and had been named 'Edomite pottery'. Since there is evidence for a Midianite origin of this ware, it should now be called 'Midianite' pottery.

The Negev ware and the Midianite pottery, for which no stratified comparisons existed at the time, could not help in dating Site 2. The 'normal' pottery could be compared with the Late Bronze Age – Early Iron Age I pottery of Palestine, but not enough identifiable pottery was found on the surface, especially none of the distinctive cooking pots of these periods, to help towards a definitive date for Site 2. Furthermore, as all sherds were surface finds, there was no way of telling whether the three kinds of pottery represented three different periods of copper workings at Timna or three different ethnic factors, operating at one and the same time. There existed, of course, other possibilities, but it became evident that only systematic excavation could solve both the chronological and the metallurgical problems.

Excavations at Site 2 1964–66

Altogether ten separate areas were excavated, chosen as those likely to answer the questions which had arisen out of the survey of the Timna smelting sites, the discovery of several apparently contemporary sites in the Arabah, and the previous detailed exploration and soundings at Site 2 itself. A number of additional soundings were also undertaken to establish the nature of several tumuli in the immediate vicinity of Site 2. These turned out to be burials, already plundered in ancient times.

Fig. 16

The immediate objectives of the excavations were: (1) To locate and excavate metallurgical installations and materials, to enable the ancient processes to be reconstructed. (2) To find workshops, stores and other structures likely to help form a picture of the organization of a large scale copper producing plant of the late second millennium BC. (3) The establishment of a reliable stratigraphy for Site 2 in order to solve the chronological problems of the relevant Timna and Arabah sites. (4) To secure sufficient stratified pottery, and other artifacts of cultural significance, for a reconstruction of the cultural and, if possible, social background of the Timna smelters and the people concerned.

Accordingly, the areas chosen for excavation fell into several characteristic groups: C, E, and G, in the immediate vicinity of large slag heaps; B1, I and K with metallurgical activities other than smelting; D–K and B2, likely sites of workshops; M was thought to

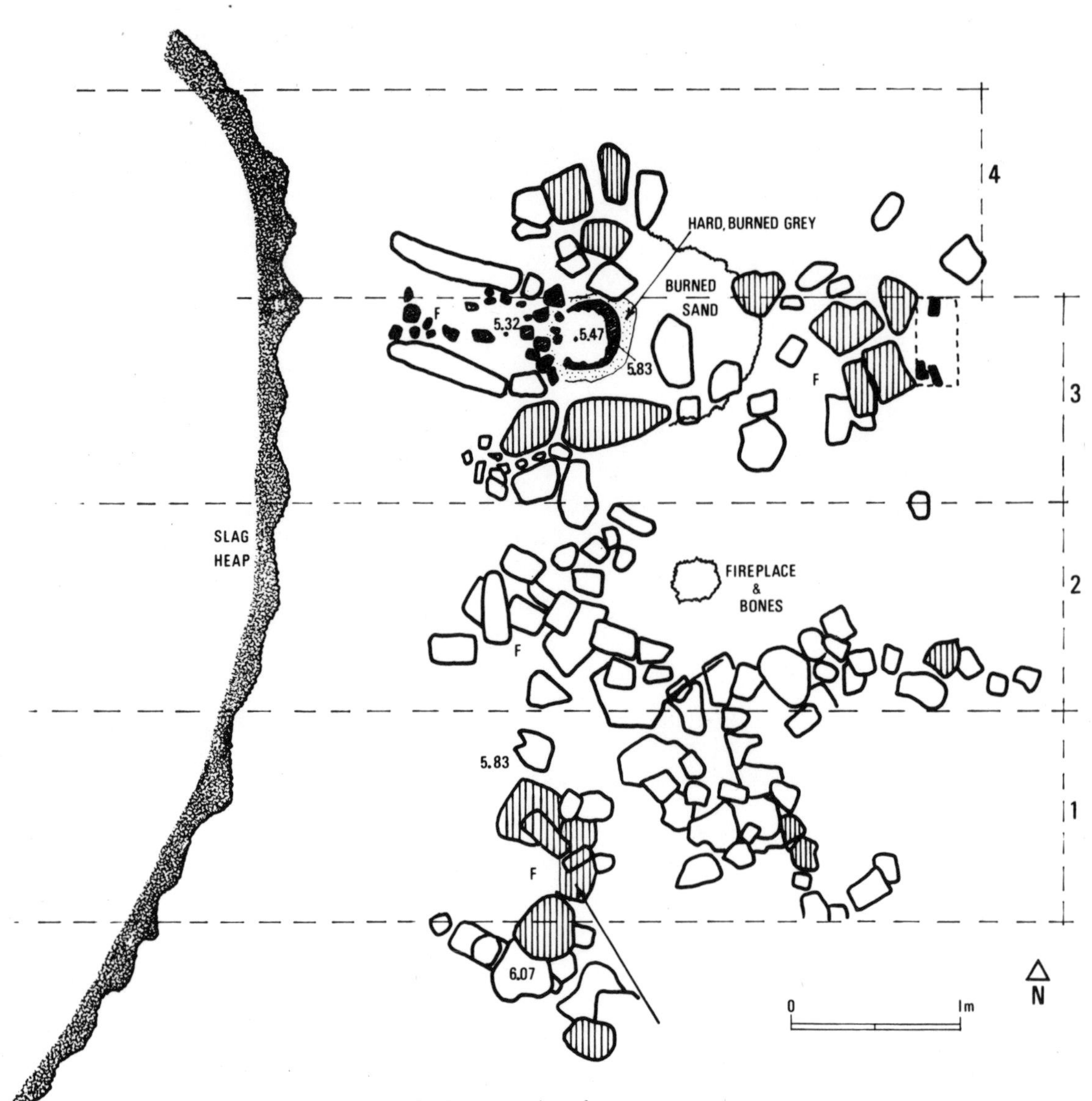

18 Site 2, Area C: Copper smelting furnaces

be a casting installation (but was, in fact, a highly interesting burial), and A and F, two isolated spots near the smelting camp, were chosen for reasons explained in the next chapter.

Smelting furnaces, Area C

Fig. 18

The remains uncovered at Area C are typical for the copper smelting installations of the Ramesside period in Timna and will therefore be described in some detail. The area excavated covers 10 × 6 m., including part of a slag heap, consisting of a 50 cm. thick, solid mass of dark, heavy, semi-circular slag. Small, porous, brown-grey slag pieces were also found in this heap and all over Area C. The slag heap is partly piled up on the slope, partly continuing also into the wadi bed. About 1 m. from the thinning out east edge of the slag heap, various ruined metallurgical installations came to light. In fact, two copper smelting furnaces, III and IV, were found, of which Furnace III was in fragmentary condition but IV was in an excellent state of preservation. Apparently parts of Furnace III were removed for further use at a nearby new furnace construction, perhaps for Furnace IV. Like all smelting installations found in Timna from the second millennium onwards, Furnace IV was located on sloping ground. A hollow, measuring approximately 1 × 2 m., was dug into the red sand layers partly penetrating into the red sandstone below. On the higher side of the slope, the depth of the hollow is about 50 cm., whilst at the other end it rises gradually to ground level. The actual smelting bowl, 40 cm. deep and 45 cm. in diameter, was located at the deep end of the hollow. No stones were used to strengthen its walls, it was simply a hole in the ground with a thick layer of clay mortar as its wall and bottom. A thick layer of slag was found adhering to the furnace walls, extending from the upper rim down to about 15 cm. above the furnace bottom. Here the slag lining was missing and seemed to have been intentionally removed, probably whilst clearing out the furnace at the conclusion of the smelting process. No slag was adhering to the well-preserved, hard-burnt, grey furnace bottom; the front of the furnace, facing west, was also missing. Two flanking stones, about 80 cm. long, formed a compartment in front of the opening which protected a shallow pit, 70 × 100 cm., dug to a level of about 15 cm. below the furnace bottom. The bottom of this pit, like the furnace bottom, was burnt light grey, and must have been the consequence of very high temperatures.

Fig. 19
Plates 35, VII

When the first smelting furnace was excavated, the excavators did not recognize this pit in front of the furnace. All that was seen was a smelting bowl with heavily slagged walls. In trying to understand the method of smelting in this furnace it was impossible to fit the large circular slag plates with a hole in their centre into the picture. Most of the slag circles were greater in diameter than the smelting bowl and no technologically sound explanation was apparent for the central hole. Careful study of the slag itself revealed sand and small stones sticking to the underside which could not have come from a clay-lined furnace bottom, and the rim of the slag was 'as-cast' over most of its surface. Invariably a small fraction of the slag rim was found to be broken off and eventually one slag cake

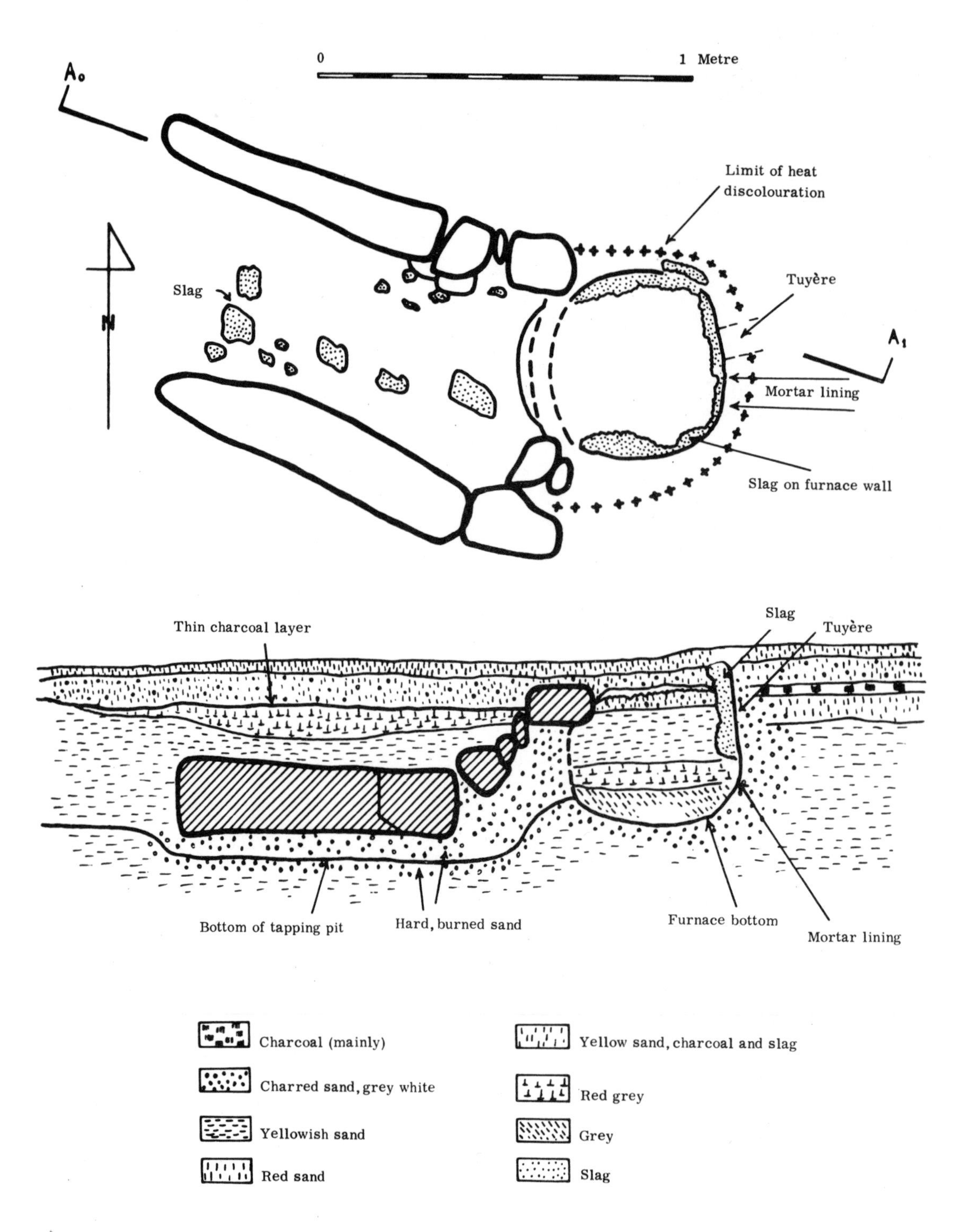

19 Plan and section of copper smelting Furnace IV

was found with a 25 cm. long 'runner' of slag attached to it. It therefore became evident that the slag had solidified outside the furnace, *i.e.*, it must have been tapped into a slag pit. The shallow pit between the flanking stones was obviously a tapping pit. The difference in height between the higher furnace bottom and the lower tapping pit guaranteed the fast flow of the hot, liquid slag; a tapping hole drilled into the side of the furnace, approximately 10 cm. above the actual furnace bottom, made sure that the metallic copper did not escape together with the slag.

In the furnaces of Site 2 there was no archaeological evidence for an intentional 'cast-in' of the centre hole in the slag and it must be assumed that it became obliterated or was not recognized by the excavators. Clear evidence, however, for such a device was found in the Roman smelting furnaces, excavated in 1969 near Beer Ora. As will be shown in Chapter VII below, the Roman furnaces are based on a similar principle to that of the furnaces of the Ramesside period in Timna and can, therefore, serve as a reliable technological comparison. It seems certain that the centre hole was cast-in to permit fast removal of the slag from the tapping pit by means of a hook.

Inside the upper half of the furnace wall, opposite the tapping hole, fragments of a clay tube were found, which must have served as a tuyère for the bellows. The aperture was about 10 cm. in diameter and almost all the clay protectors of the bellows end, found in great numbers around the furnaces of Site 2, fitted it quite well. As there was only one tuyère in the back wall of the furnace and none at the sides, there could not have been more than two bellows to a furnace. In fact, the two types of clay protectors found indicate the use of one bellows at the back, with the tuyère at an angle of about 70° to the horizontal, directed towards the centre of the smelting bowl. In this case the clay end of the bellows tube would be pierced by a straight running hole. A second bellows seems to have been operated through the front wall, above the tapping hole. Here, a tuyère went horizontally through the furnace wall and the clay protector was pierced by a hole bent at its end to direct the airflow, as required by the smelting process, to the lower half of the furnace. This second tuyère was not actually found in any of the excavated furnaces, but the reconstruction, based on the evidence of the numerous tuyère ends present with purposefully bent air holes, seems technologically sound.

Fig. 20
Plates 32–34

Around the upper rim of the furnace, which must have projected above ground, several large, flat stones formed a working platform. All the stones used for Furnace IV, as most of those used for furnaces

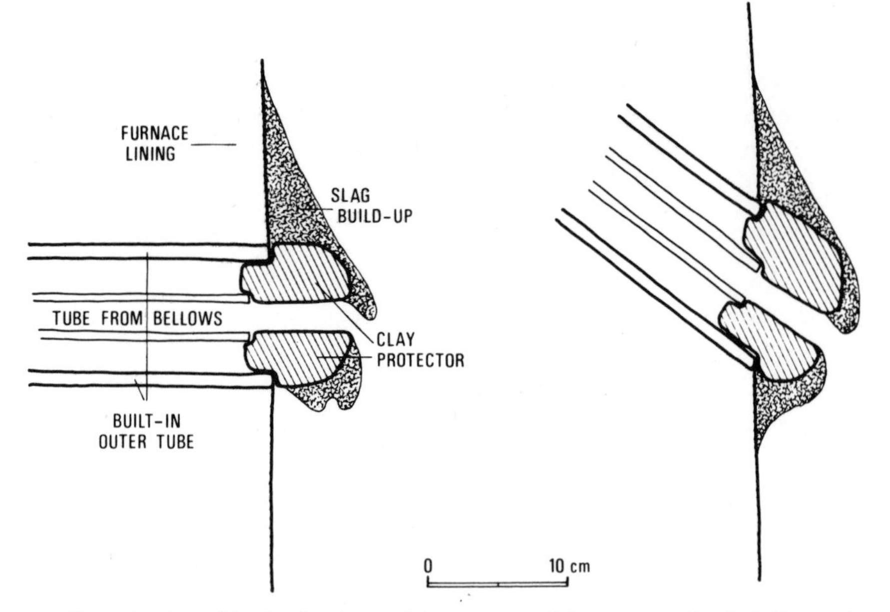

20 Reconstructions of details of tuyères and the two types of clay protectors for the bellows ends

in Timna, were of dolomite, which is the most fire resistant stone in the area. Stratigraphical evidence showed that Furnace IV was dug through two earlier working surfaces, indicated by a change of colour and material, with some metallurgical waste and a thin charcoal layer.

Area C showed everywhere three working-floors, one on top of the other, and evidence of intensive smelting activities in the form of several destroyed and dismantled furnaces. The three working furnaces made up one industrial occupation layer of about 60 cm., above undisturbed red sand and water-laid pebbles. In between the working seasons, most likely interrupted by the hot summer months, the surfaces collected a thin layer of wind-borne, yellowish drift sand. This is evidence for seasonal activities at the site, yet the pottery on all the working-floors of Area C belongs to the three kinds described above, found everywhere mixed together. There is no difference whatsoever between the sherds from the different working floors.

AREA G
Plates 36, 37; *Fig. 21*

Area G, located next to a large slag heap, was first excavated as a square of 4 × 4 m. in 1964 and a smelting furnace, Furnace II, unearthed. This furnace is, in principle, the same as Furnace IV except that it is much stronger built. Instead of just a hole in the ground, a semi-circular stone wall of maritime cambrian dolomite was meticulously built into the deep end of the furnace hollow, with its open side above the tapping pit, protected by two long flanking stones. As in Furnace IV, the wall and bottom were lined with a thick layer of clay mortar and a clay tuyère was found going through its back wall. Around the furnace was a solid working-platform of large, flat stones.

In 1966 Area G was enlarged to 7 × 7 m. and excavated to bedrock. Five superimposed metallurgical working-floors were carefully peeled off, and metallurgical materials and waste of each floor taken

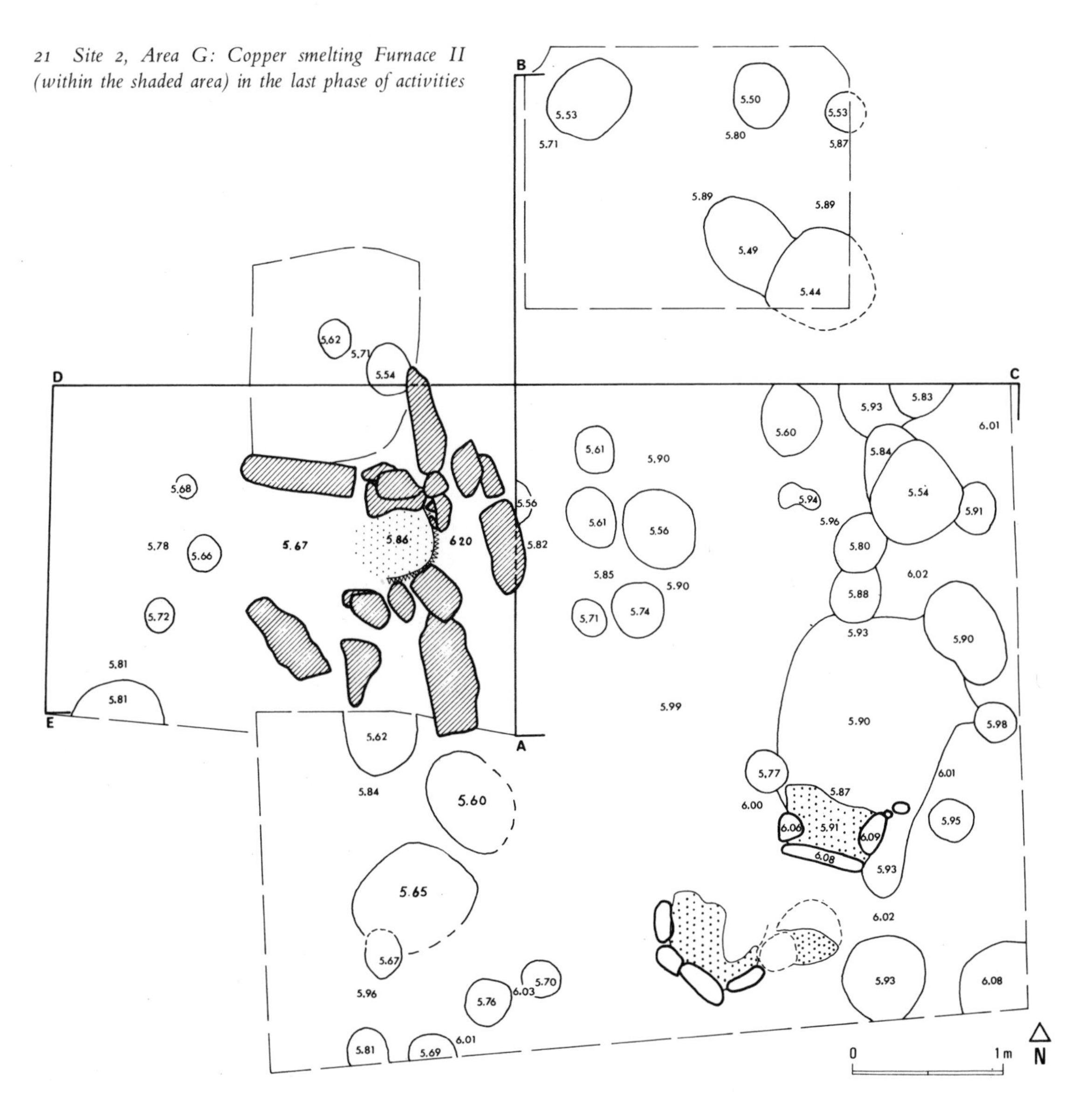

21 Site 2, Area G: Copper smelting Furnace II (within the shaded area) in the last phase of activities

for separate analysis. Furnace II belongs to the last phase of metallurigical activities at Site 2, with its working-platform actually protruding above the present ground level. The lowest working floor rested on bedrock and consisted of the remains of completely dismantled smelting furnaces, including two furnace bottoms *in situ* and numerous small slag pieces. Several small, mostly bell-shaped, pits 20–40 cm. deep were found in Area G, dug into every working-floor. In the lowest floor thirty-six pits were counted, many cut directly through into the bedrock, others were dug down from one of the working-floors above, penetrating through one or several floors. The pits, used for storage, were common to all areas of Site 2 and seem to be characteristic of metallurgical sites of this period. Inside the carefully cleared thirty-six pits, a variety of materials was found. It was mainly metallurgical waste, charcoal, slag and burnt sand, which had gradually filled the empty pits. A number of the pits, however, were found to contain some of their original contents, and were highly informative: there were date kernels, many broken bones of fully grown goats, some cattle bones and some bones and teeth of donkeys. A large quantity of ostrich eggshells was also collected. There were also many beads, lumps of copper ore, fluxes and some of the pits next to the furnace opening contained a solid mass of charcoal and donkey dung, apparently for the smelting fire. On each working-floor as well as in many of the pits, found sealed by a floor above, all three kinds of pottery were present.

In Area E an additional smelting furnace, Furnace I, was found next to a large slag heap. This furnace was visible above ground prior to excavation, simply hidden underneath some debris and was, in fact, the first furnace found by the Arabah Expedition in 1962. It was also stone-built, having two large flanking stones along its tapping pit, and several pits were found around the furnace, some full of charcoal. The working-floor of Furnace I was at such a high level, on the higher slope of the valley, that only several centimetres of recent wind-borne sand had to be brushed away to show the outline of the storage pits. The pottery found on the surface and inside some of the pits was the same as everywhere else on Site 2.

Crucible melting furnaces

In Area B1, square E6, on the east side of Site 2, a large area of the present surface showed a distinct dark discolouration, with much small, porous slag and charcoal bits mixed with red sand, overlaying a thin layer of yellow wind-borne driftsand. A trial trench dug across this dark area, which altogether measured about 10 × 7·5 m., revealed

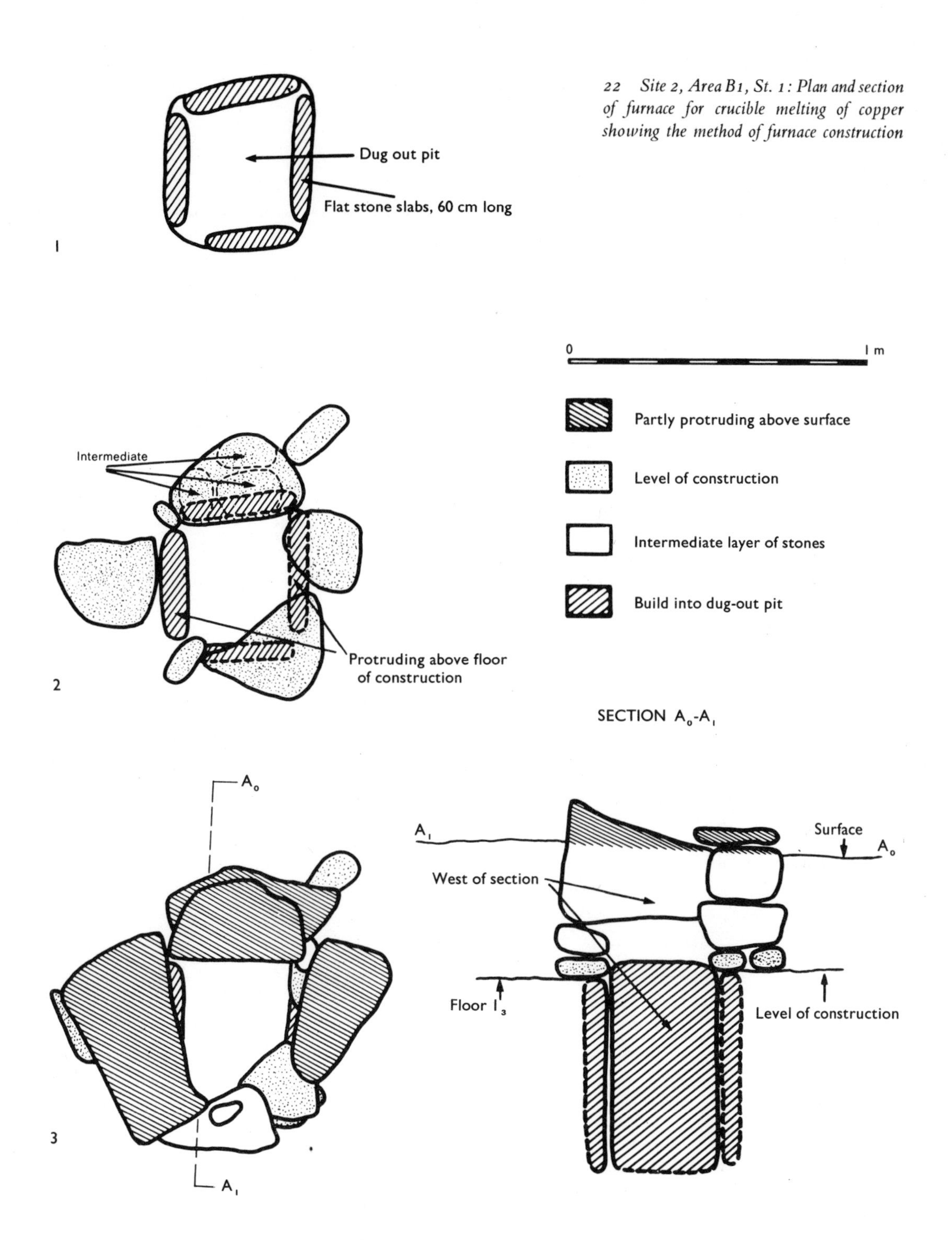

22 Site 2, Area B1, St. 1: Plan and section of furnace for crucible melting of copper showing the method of furnace construction

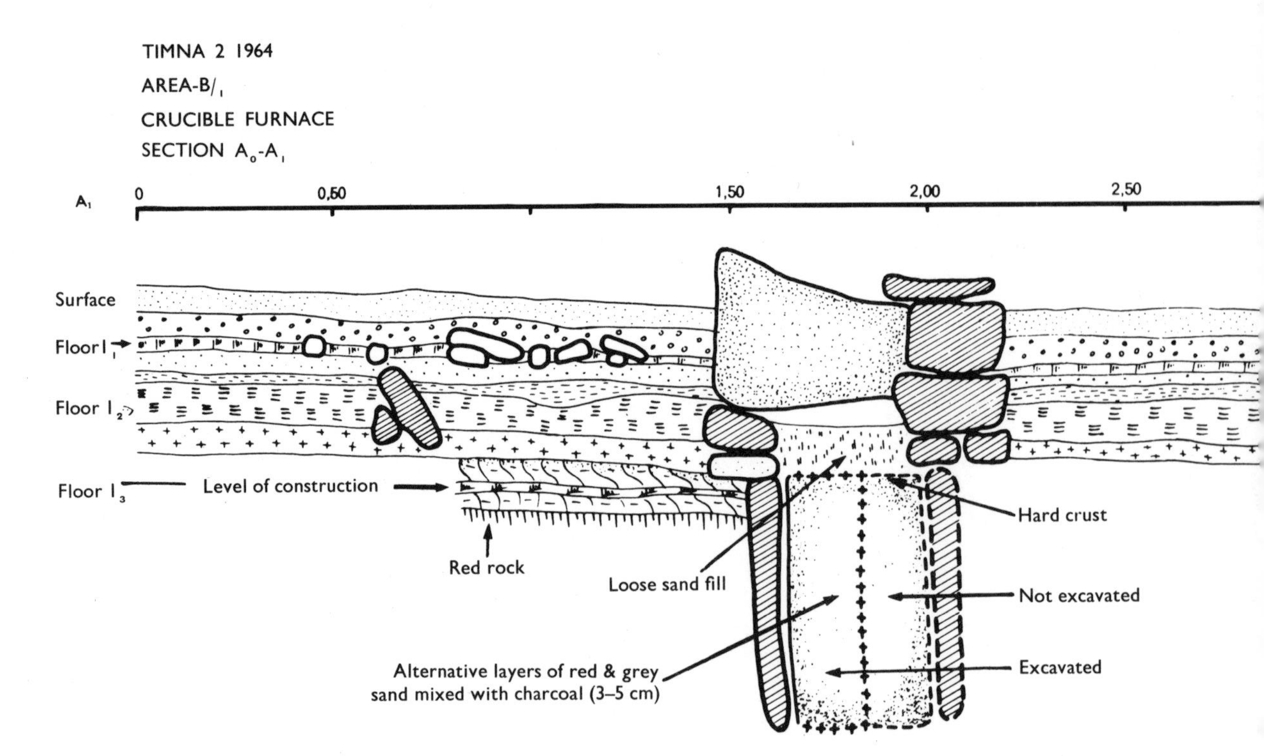

23 Site 2, Area B1, St. 1: Section of crucible furnace from casting workshop

an oval-shaped, solid central core of charcoal, about 3·5 × 2·5 m. and 30 cm. thick, with charcoal and charcoal dust spreading out over the surrounding ground. As the charcoal also contained a considerable quantity of unburnt wood, mainly branches of acacia trees which still grow in the Arabah, we consider this to be a 'charcoal pile', used to manufacture charcoal for the furnaces of Site 2.

Plate 38 *Fig. 22*

Immediately to the south of the charcoal pile, a square stone compartment, St. I, measuring 40 × 40 cm. with a depth of 80 cm., was excavated. It was in fact a square pit, lined by flat stone slabs and its top was covered with several large stones, leaving a narrow opening at its south side. The installation was found filled with charcoal ash, pieces of charcoal and small bits of slag – it was a hearth used for the crucible melting of copper. Pieces of metallic copper, probably drops and prills of copper, extracted from smelting slag, were melted in small crucibles inside this hearth and cast into moulds.

24 Melting crucible found at Site 2 (Plate 41)

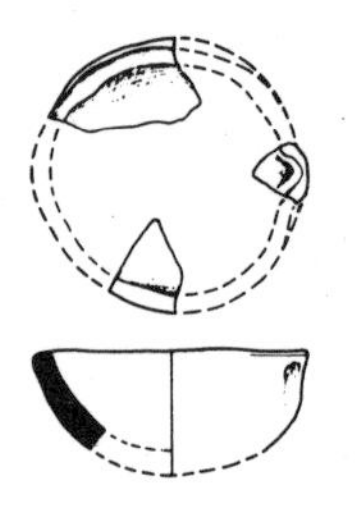

In Area I another installation, built of flat stones, was found to contain a quantity of black ash and burnt material. Also, a complete,

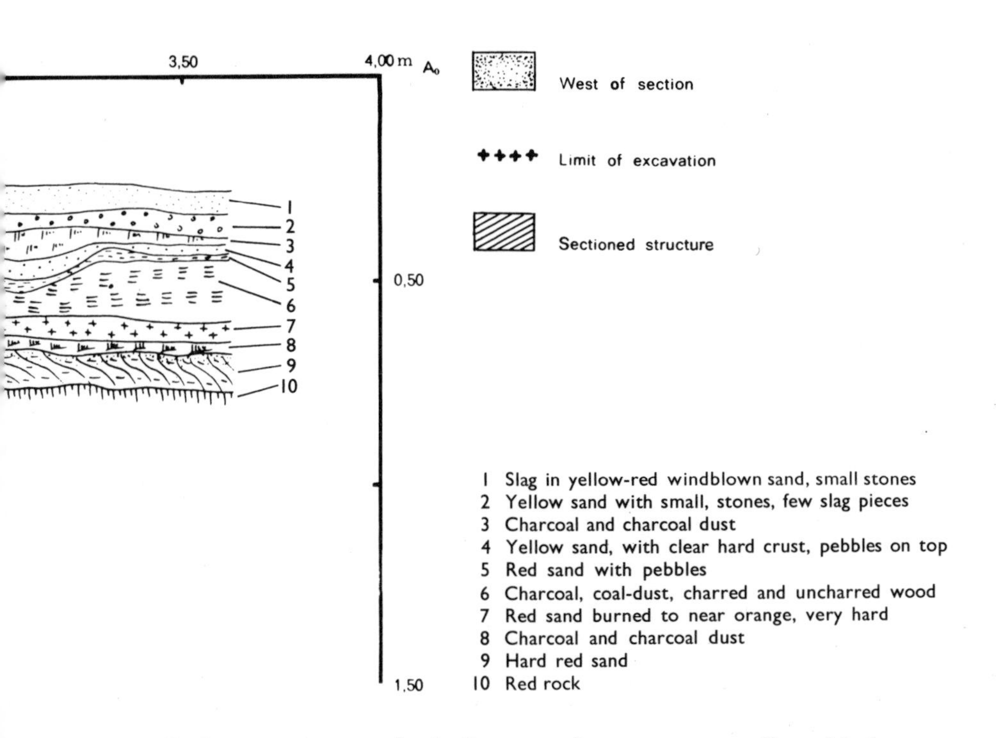

small clay-protector of a bellows and a copper needle with its eye still intact were found nearby. Area I was thus also the site of melting and casting with a crucible-melting hearth operating next to it. A further casting site with several crucible-melting furnaces was found in Area K.

COMPLETE INDUSTRIAL UNITS IN AREAS D–K AND B

Area D–K is located at the northern end of Site 2 and the building complex unearthed in this area must have been connected with the furnaces and the slag heap of Area C and perhaps also with Furnace II.

During the first season of excavations, in 1964, a two-roomed building was found, measuring 4 × 8 m., and was thought to be a storehouse. In the second season, in 1966, the area of excavation was considerably extended and reached over 400 sq. m. This complete area was taken up by the large building complex of a complete working and storage unit (D–K).

Fig. 25

The centre of this structure is a courtyard, 8 × 11 m., with a very large, stone-lined storage pit for ores (Locus 1013). The stone lining, of very heavy large stone slabs, had slipped into the pit and it was therefore only excavated to a depth of approximately a metre,

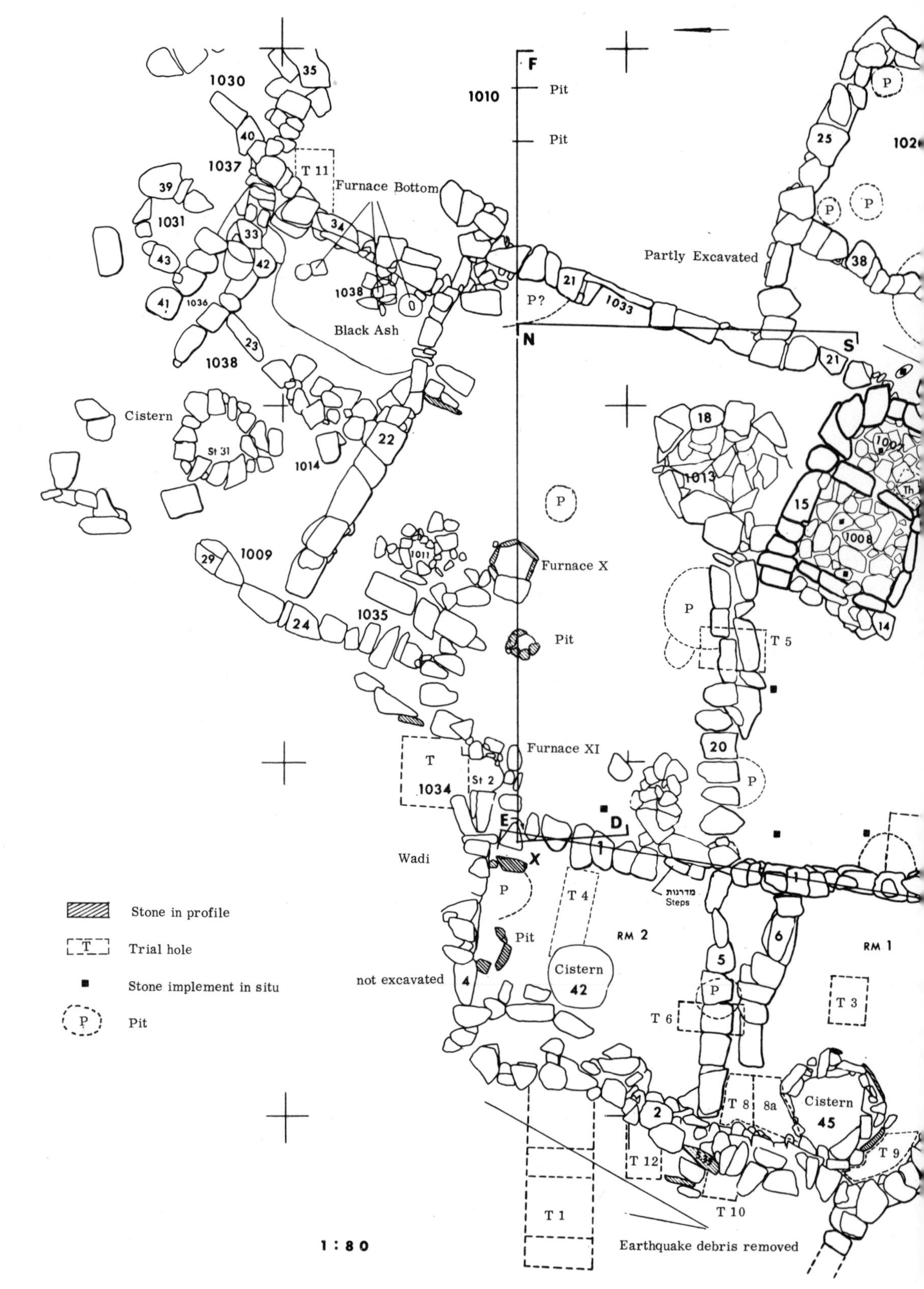
F
1010
Pit
Pit
1030
35
40
1037
T 11
Furnace Bottom
39
1031
34
33
42
43
1038
41
1036
Black Ash
23
1038
P?
21
1033
Partly Excavated
25
38
N
S
21
Cistern
St 31
22
1014
18
1013
15
1008
P
Furnace X
1009
29
1011
1035
24
Pit
P
T 5
14
Furnace XI
T
1034
St 2
20
P
E
D
Wadi
X
1
1
מדרגות
Steps
P
T 4
Stone in profile
Pit
RM 2
6
RM 1
Trial hole
5
Cistern
42
not excavated
4
Stone implement in situ
P
T 6
T 3
Pit
T 8
8a
Cistern
45
2
T 12
T 9
T 1
T 10
1:80
Earthquake debris removed

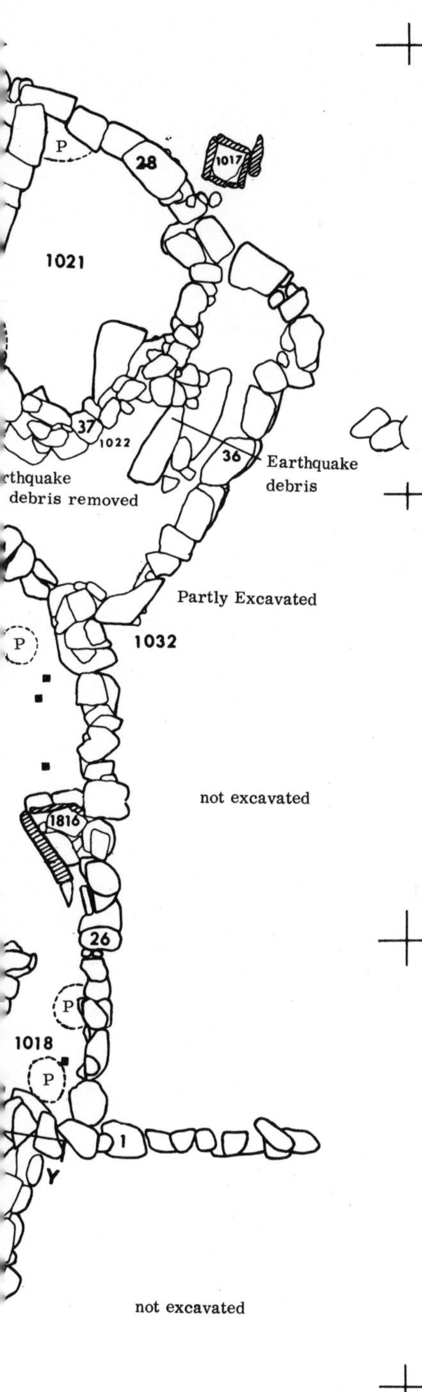

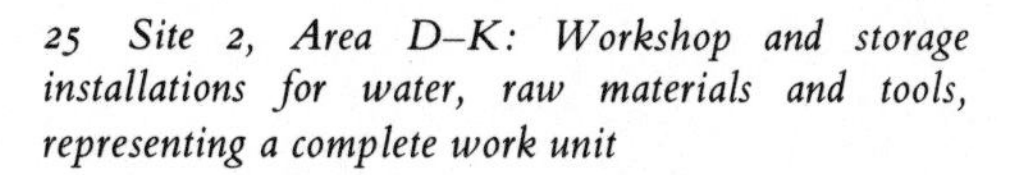

25 Site 2, Area D–K: Workshop and storage installations for water, raw materials and tools, representing a complete work unit

Plate 29

though the pit doubtlessly must have been much deeper. Next to it a unique stone platform (Locus 1007–1008) was found with many crushing and grinding tools and crushed copper ore bits *in situ*. It appears that the courtyard was the centre for the preparation of the smelting charge for the smelting installations nearby. Many ore-grinding tools were found *in situ* standing in all corners of the courtyard.

Donkey dung was found in a heap just north of the ore pit and a line of droppings ran across the courtyard towards the entrance at the north-west corner, near Locus 1034. These must have been the droppings of the pack-animals which brought the ore from the copper mines.

Solid buildings, mainly of limestone, were joined to the outside walls of each side of the courtyard. The walls were dry built and the method of construction, especially at the corners, was very reminiscent of the head-and-stretcher building method of the Early Iron Age.

On the west side in Area D two rooms (RM1, RM2) were each found to have an underground water container, and RM2 also had a line of storage pits along its northern wall. The pits contained a store of stone implements and ores as well as some light yellowish clay which was used, apparently, for the preparation of tuyères, clay protectors for bellows, furnace and cistern linings and perhaps also pottery. The water containers are solid circular structures of stone, built into pits dug into the red sand and sandstone and made water-tight by a clay mortar. The upper rim of Cistern 42 was 100 cm. in diameter for a depth of 50 cm., then it suddenly widened out to 1·70 m. with a depth of 1·33 m. Cistern 45 was a simple stone and mortar lined pit, 1·25 m. deep and 1·10 m. in diameter.

Plate 28

The two cisterns in RM1 and RM2 were found filled with debris and metallurgical waste and there were many interesting finds, including pottery. Both cisterns belonged to the very first stage of activity in this area and seem to have fallen out of use before being left to fill up. Perhaps the filling-in of these disused water holes was done on purpose, as some large stones were also found with the debris. Yet, at this later stage the 'cisterns' may also have served for the storage of ore. It is clear that water was carried to them in containers for cool storage and that they were not filled by run-off rain water.

On the east side a two-roomed structure, 6·5 × 5 m., was uncovered (Loci 1020, 1021). Although it was used as a workshop, it could also have served as living quarters. A small cooking stove

26 Close to Site 2 was found a carefully built corbel-vaulted tomb (*Figs. 27, 28*). It had contained the remains of two individuals of Afro-Egyptian origin. Only one skull was present and this had been deliberately laid on a stone head-rest

27 Aerial view of smelting Site 2, now called the 'Mushroom Camp' from the curious natural formation nearby. In the background is the Wadi Arabah and on the hill to the left of the 'mushroom' is Area F, the site of a High Place

28, 29 The industrial Area D-K. *Above*, a stone-built cistern in one of the rooms of the workshop. → *Below*, a view of the centre courtyard with a stone platform where ore was crushed. Behind it is a storage pit for the ore. Round about are storage rooms and a casting workshop in the background

← 30 Area B1, a workshop with a semi-circular crushing platform

← 31 Fragments of circular tapping slag

32-34 Three views of a clay protector for a bellows end (*Fig. 20*). *Left*, the interior side, which is fitted into the bellows tube, is free of slag; *centre*, the side view shows the slag dripping down and solidifying, and, *right*, the external side, facing into the furnace, shows the slag build-up

35 Smelting Furnace IV (*Fig. 19*), showing the smelting bowl with its slagged walls and, before it, the tapping pit between two flanking stones

← 36, 37 The stone-built Furnace II has a slag pit in front of it. The detail, *below*, of its mortar-lined smelting bowl shows the remains of a tuyère in the back wall

38-41 Casting furnaces and crucibles in Site 2. *Right*, crucible furnace St. 1 (*Figs. 22, 23*). *Below left*, crucible melting furnace X in Area D-K (*cf. Fig. 25*); *right*, a lump of crude copper retaining the shape of the crucible in which it was melted, similar to the example shown (*Fig. 24*)

42-44 Wheel-made bowls (*above and centre; cf. Figs. 30, 7, 8*), and a hand-made Negev-type cooking pot (*below; cf. Fig. 31, 5*)

45 Base of a Negev-type cooking pot showing mat-impressions

46, 47 Ramesside steatite scarab. The engraved base shows a sphinx walking to the right preceded by the feather (the sign for 'truth') and the sacred uraeus. Above his back the two signs read 'The Beneficent God'

48–52 Midianite pottery from Smelting Camp 2 (*cf. Fig. 32*). It was brought to the site from northern Arabia. Most of the pottery is decorated with sophisticated bichrome designs

53, 54 A Midianite jug and bowl decorated in bichrome style. The jug (*Fig. 35*, 1) was found at Site 198, a funerary shrine on top of 'King Solomon's Pillars', and the bowl comes from Smelting Camp 2 (*Fig. 32*, 2)

55-58 Rock-engraving 2 in the Ramesside copper mines (*Fig. 38*). The details (*opposite*) show Egyptian soldiers in ox-drawn chariots brandishing New Kingdom-style battle axes, and straight-horned oryx and their Midianite hunters. *Left*, white sandstone votive basins found beneath the rock-engravings, similar to those found in the Hathor Temple

59, 60 Rock-engraving 1, found on the cliff-face just above the standing figure, and the remains of a damaged, rock-cut water cistern to the left. *Below*, a detail of the engraving (*Fig. 36*)

61, 62 The High Place at the northern end → of Site 34 (*see also* Plate 19) is a natural outcrop. On its summit stands a rock altar (*below*), surrounded by a series of cup-marks. They were probably used as libation bowls of a *bamah* which served the Semitic workers at the smelting site

63, 64 Jezirat Fara'un (or 'Pharaoh's Island') appears to have been the mining port connected with the Ramesside mining expeditions to the Arabah ('Atika'). It was later used by the Israelite Kings (as Ezion Geber), then by Nabataeans, Romans and Mamelukes

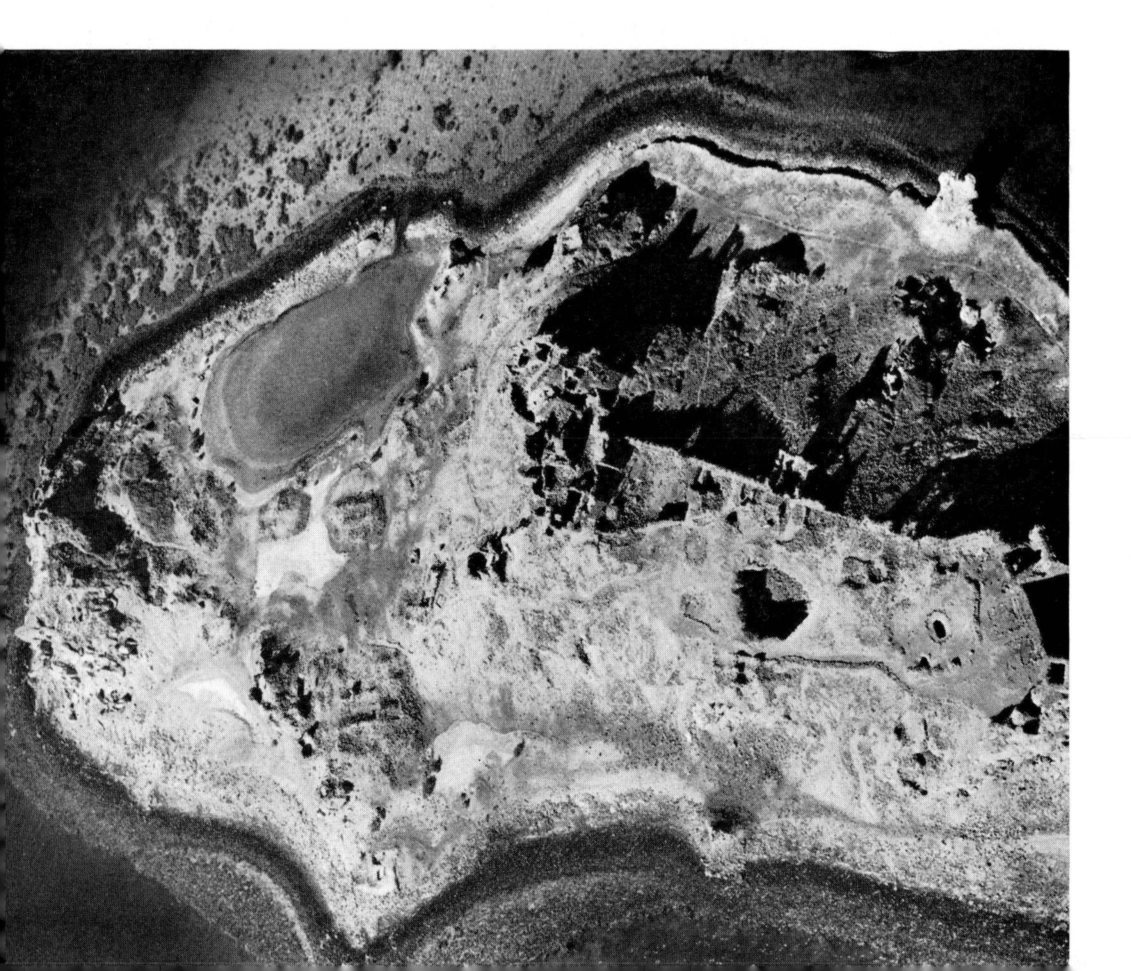

(Locus 1017), built of stone slabs, was found outside this structure. More rooms seem to have been attached also at the south end of D–K but this area was not fully excavated.

The north end of D–K was a casting workshop, with well-built crucible-melting furnaces (Furnace X, Locus 1011) standing inside the actual courtyard. The building attached here to the outside of the northern courtyard wall showed several phases of construction and repairs, and altogether five working floors could be distinguished. On each of these floors remains of casting furnaces were found, often only as burnt hard furnace bottoms still *in situ*, and also very much wood ash, charcoal dust, charcoal pieces, copper pellets and prills, slag and slagged crucible fragments. An additional 'cistern' (St. 31) was found here, 1·87 m. deep.

Plate 39

Plates 40, 41

With storage of raw materials, ore dressing and the preparation of the smelting charge in workshop D–K, the actual smelting of the copper took place in the smelting furnaces nearby. As some of the product remained in the slag as small drops and prills of metallic copper, it was worth while to collect some by breaking up the slag. The copper bits thus extracted seem to have been smelted and cast into simple implements, ingots or even votive gifts for the nearby Hathor Temple. Areas D–K and C represent, therefore, the full cycle of copper production of Ramesside Timna.

Area B2–3, together with the crucible-melting furnace of Area B1, described above, was another industrial unit. Here, an approximately 5 × 4 m. structure was uncovered with a deep, stone-lined ore pit and a solid, semi-circular crushing platform. The ore pit was excavated to a depth of 1·15 m. Stone crushing tools and a lot of finely crushed copper ore was found *in situ* on the working platform. A small domestic fireplace with many crushed bones of goats was found in the vacant corner of the workshop, which also seems to have served for habitation.

Fig. 26

Plate 30

The walls of workshop B2–3 were very solid and the ground carefully levelled by cutting a step into it on the east side of the structure, which is here located higher up the slope. To make sure that the soft sand, laid bare by this cut into the slope, would not collapse into the workroom, an ingenious method of stone casing was devised with the stones standing on their side inside the freshly cut step, thus serving at the same time as foundations for the wall above. The lowest stone layer of the wall itself was partly resting on this line of casing stones and partly on the slope behind.

The construction method of levelling sloping ground by cutting steps at the right spot and encasing them with flat stones, has been

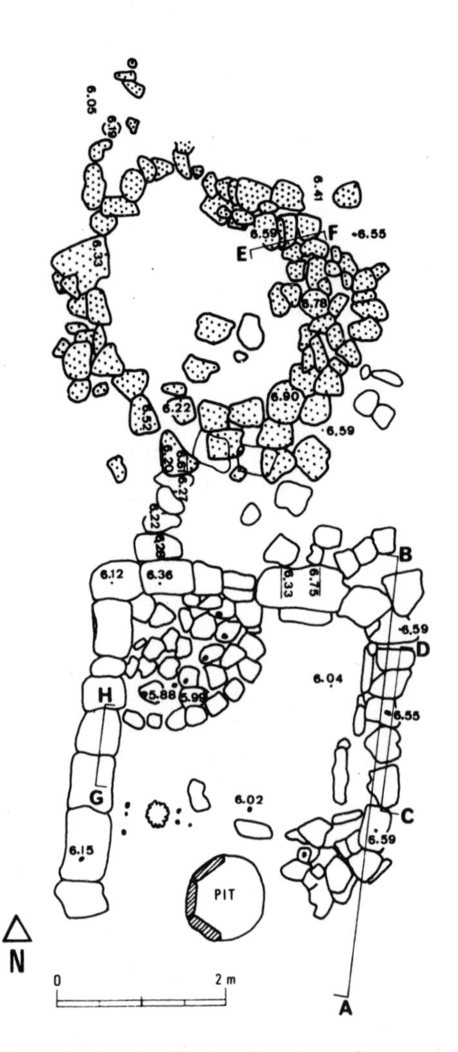

26 *Site 2, Area B2–3: Ramesside ore-crushing installation. The upper, round structure belongs to the last phase of activity at the site, apparently by the local Midianites*

found also in K (wall 20) and seems to have been standard practice in this period. Complementary to this, on the lower side of the slope, which would drop even below the level attained by the step-cutting opposite, a flat and straight working floor was created by filling in the missing ground up to the required level. Wall 2 of D, situated at the very end of a slope but somewhat above the wadi bed, was investigated by trial-hole 8, dug against it from inside. It was quite evident that the wall here was solidly built to its maximum depth, actually right into the wadi bed and down to bedrock, and the space between wall and slope was filled in with a packed mass of sand.

Fig. 25

Another characteristic of the Ramesside construction method in Timna is the frequent use of small, bell-shaped, storage pits, with or without stone lining, found everywhere in the floors of the buildings and courtyards of Site 2.

On the present surface of Area M a group of medium sized stones was found forming a circle; others, mainly flat stones, were strangely sticking out of the ground as if especially to mark this spot. Nothing besides the small porous slag, charcoal and charred sand, was visible above ground. Yet, as this area was close to the workshops and furnaces of Area B and slag and crucible fragments were found on the surface, it was assumed that Area M was the site of melting casting installations. An explanation for the large quantities of porous, small and light slag, concentrated in this area of camp 2 was hoped for.

Stripping off the present surface layer, it immediately became obvious that the strange standing stones are not actually connected with the circular structure being uncovered, but served as 'markers' of some sort. Digging down further it became evident that here was some kind of fill because all was massive metallurgical waste thrown together without any order, mixed with sand. In the middle of this fill a structure of heavy stones came to light, closed on top by layers of flat stones to form a carefully built corbelled vault.

Fig. 27

Clearing down everything around the structure, red sandstone rock was reached after the removal of a large quantity of metallurgical waste. A shallow pit, about 4·5 m. in diameter had been dug into the red sand and a stone structure, 2·30 × 2·0 m. and about 1·0 m. high, was erected therein. Clearing out the inside a skeleton was found at its bottom, its head on a flat stone intentionally placed as a head-rest. The skeleton lay contracted on its left side, one arm under the thigh-bone and the other under the skull. The tomb contained no other objects. At the time of the excavation, it seemed that one body only had been placed in the burial chamber without any covering; the firm sand found above it was drift sand that had infiltrated through the roof. However, examination of the bones by Dr N. Haas of Tel Aviv University showed that, in fact, 'there were skeletal remains of two individuals. The first one, almost complete, is an adult male, 25–30 years old, with peculiar proto-negroid features. The second one, found without calvarium, was an adult of 18–22 years. No racial diagnosis of these remains was possible.' According to Dr Haas, the Timna skull, with its peculiar anthropological typological features, could not in any way belong to a population originating from the Syria–Palestine area, but an African origin is strongly indicated. It is similar to the proto-boscopoid type found in ancient Ethiopia and at Nagada in Egypt. It seems that the second possible comparison to the Timna skull with Early Indian proto-diavidoid types, found in the area of the Harappan culture, can be safely dismissed but should, nevertheless, be mentioned here.

Fig. 28

Plate 26

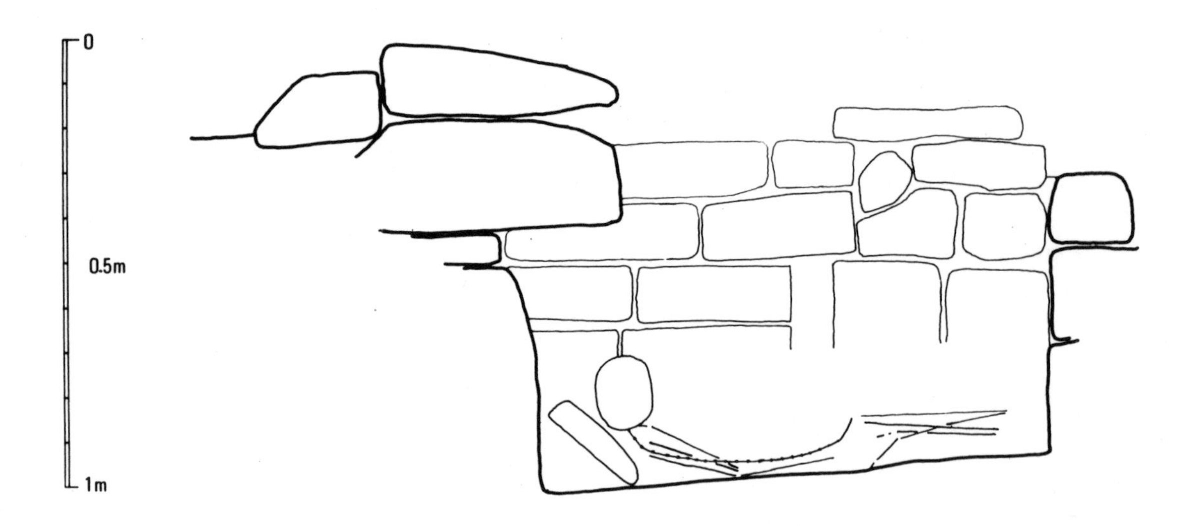

27, 28 Site 2, Area M: Section and plan of the corbel-vaulted tomb with skeleton in situ

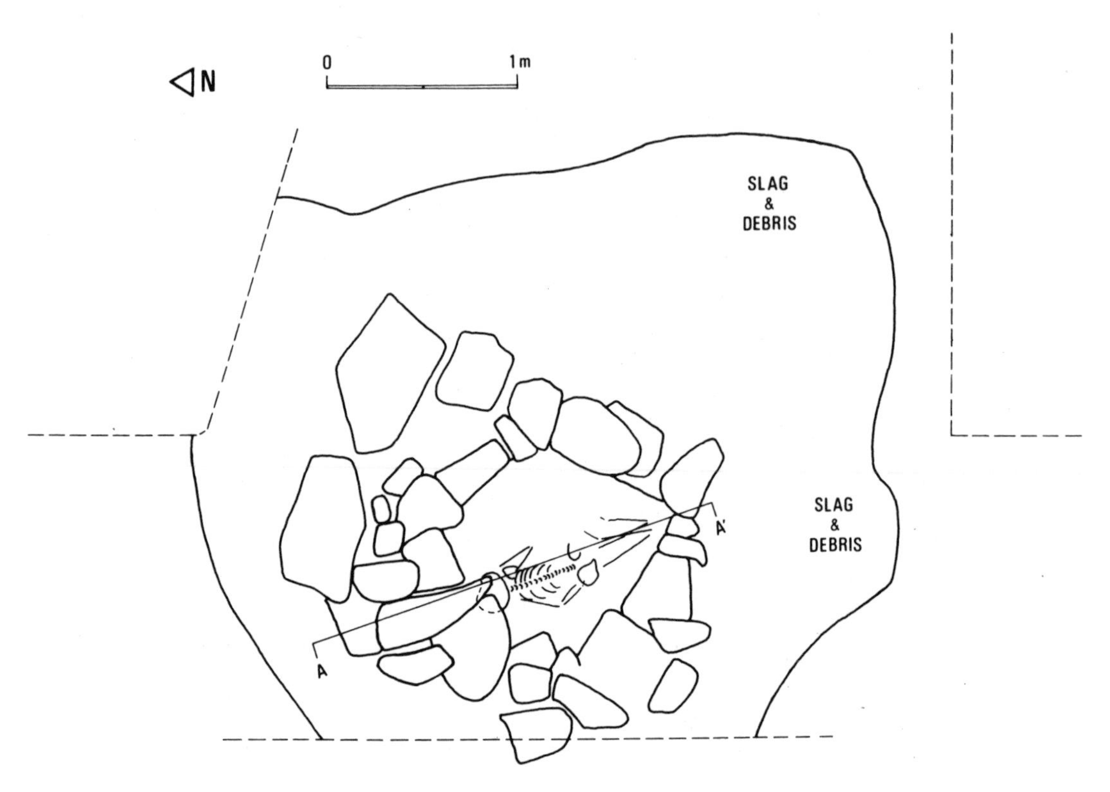

The finds and stratigraphy of Site 2

Site 2 was essentially a copper smelting camp and, besides pottery, only a few non-metallurgical finds were made during its excavation. There were, of course, very many tools, mostly the typical saddle-backed querns of varying size, made of hard gritty red sandstone. There were also very numerous flints, granite and sandstone hammer-stones, mortars and pestles.

Most of the bones found were crushed, often to small fragments, as the meat was scraped off for eating; the majority of the bones came from goats but some might have come from ibexes. There were also numerous donkey bones and the remains of Red Sea fish. It is of considerable interest that several camel bones were found in the excavation, bearing in mind the early date of the site. Whether the camel was domesticated or wild, used as food or as pack-animal remains, so far, an open question.

Fig. 29

Several copper implements were found, some showing signs of use, others were in an 'as cast' stage and quite unfinished. A toggle pin of Early Iron Age type showed a 'cast in' hole, but the hole was found clogged by bits of mould and had never been used. Among the metal finds was a heavy copper spear-butt, found in Area D, a fine knife-blade, a ring, a needle and several awls, also a hook and one arrow-head. All these copper implements seem to have been made locally. Simple beads were found in most areas of the excavation. Whether this means the presence of women amongst the workers or the use of beads by the male workers themselves, or the manufacture of votive beads which evidently took place at Area F, described in the next chapter, cannot be decided.

Plates 46, 47

A scarab was found in Area K. Made of steatite, 1·6 cm. long, 1·2 cm. wide and 0·8 cm. high, it shows a standing, human-headed, bearded, sphinx wearing an apron, which is preceded by an uraeus and *maat* (the sign for truth in the form of a feather). Over the sphinx is written: 'The Beneficent God' and underneath is the sign for 'Lord'. It is dated to the reign of Ramesses II of the XIXth Dynasty and a second scarab found some years ago on the surface of Site 2 by a casual visitor was of similar date.

Figs 30–32

A large quantity of sherds was found in the excavation of Site 2 and made the subject of a detailed study by Y. Aharoni. Compared with the rather meagre pottery from the surface collection, the excavation secured many additional types, including pieces important for dating.

Plates 42–45, 48–54

In all areas the same three kinds of pottery, collected on the surface during the previous surveys were found together on most of the different working floors and in many of the pits. The excavations

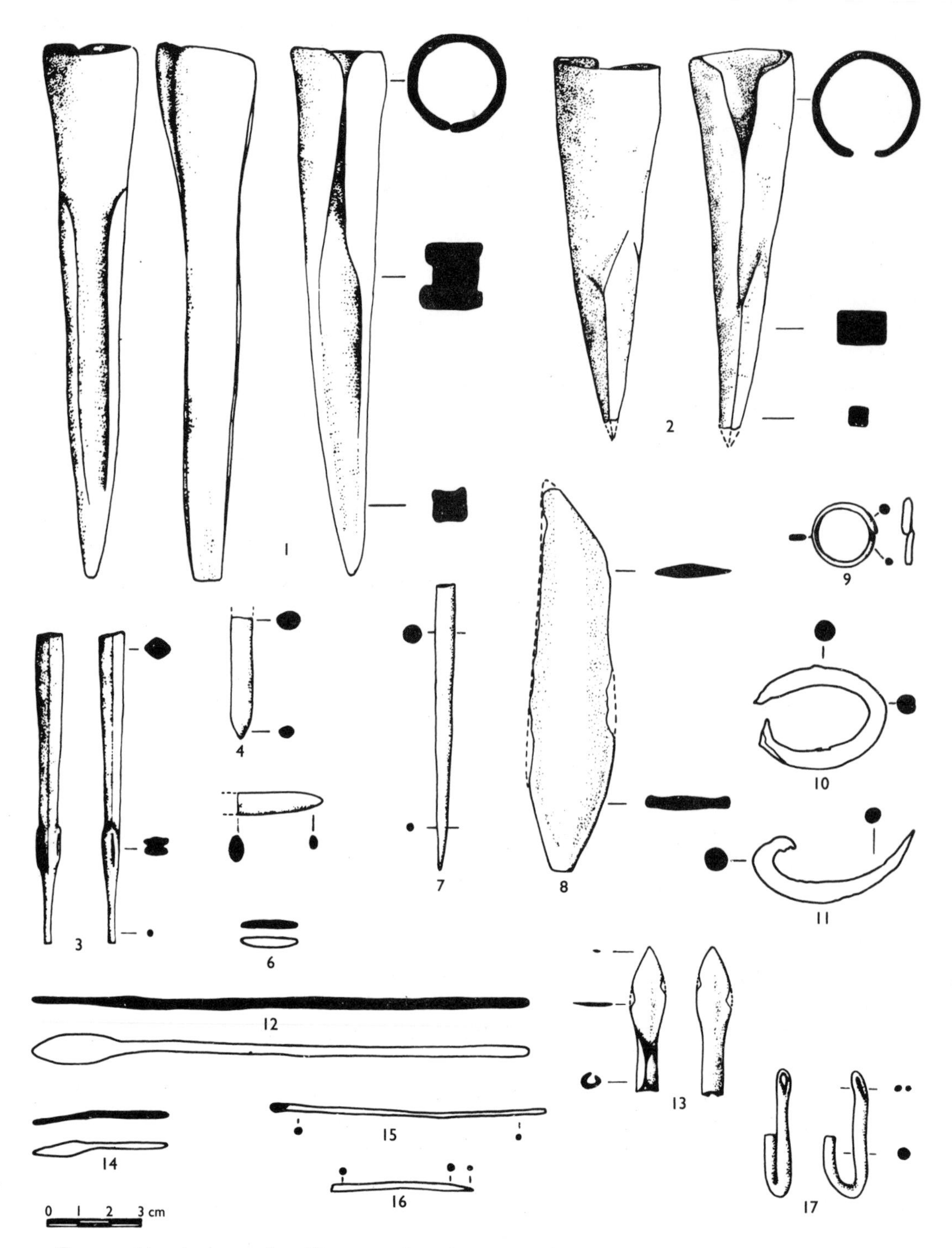

29 Copper and iron implements from Timna: 1–2 Spear butts; 3 Toggle pin; 4–5 Copper awl tips; 6 Tiny copper awl; 8 Copper knife; 9 Copper ring; 10–11 Iron bracelets; 12 Copper spatula; 13 Copper arrowhead; 14 Tiny copper spatula; 15 Copper needle; 16 Copper pin; 17 Copper hook

proved, beyond a shadow of doubt, that these three distinct kinds of pottery – 'normal' wheel-made pottery, 'Negev-type' pottery and 'Midianite' pottery – were used in Timna at one and the same time. But, important as this conclusion may be from the stratigraphic, historical and ethnographic aspects of the excavations, none but the normal, wheel-made pottery could be of any real help for the dating of the site, because only comparative material for this kind of pottery was available. Luckily, quite a number of distinctive types were secured, including cooking pots of the shallow, open and carinated type, with small, folded, triangular-shaped rims and no handles. These pots closely resemble the typical Late Bronze Age cooking pots of Palestine and of the neighbouring countries. Many large carinated bowls with handles, and storage jars with pointed and

30 Late Bronze Age-Early Iron Age I wheel-made pottery from Site 2

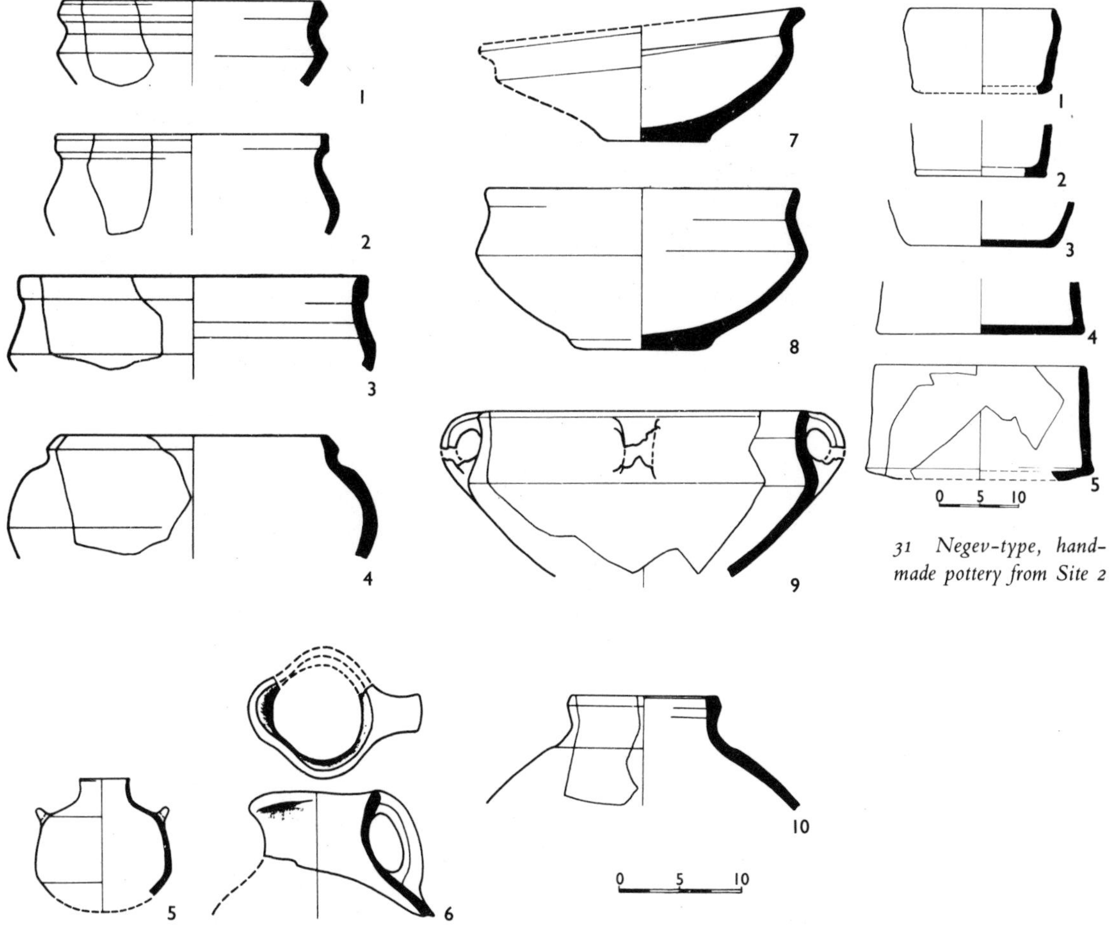

31 Negev-type, hand-made pottery from Site 2

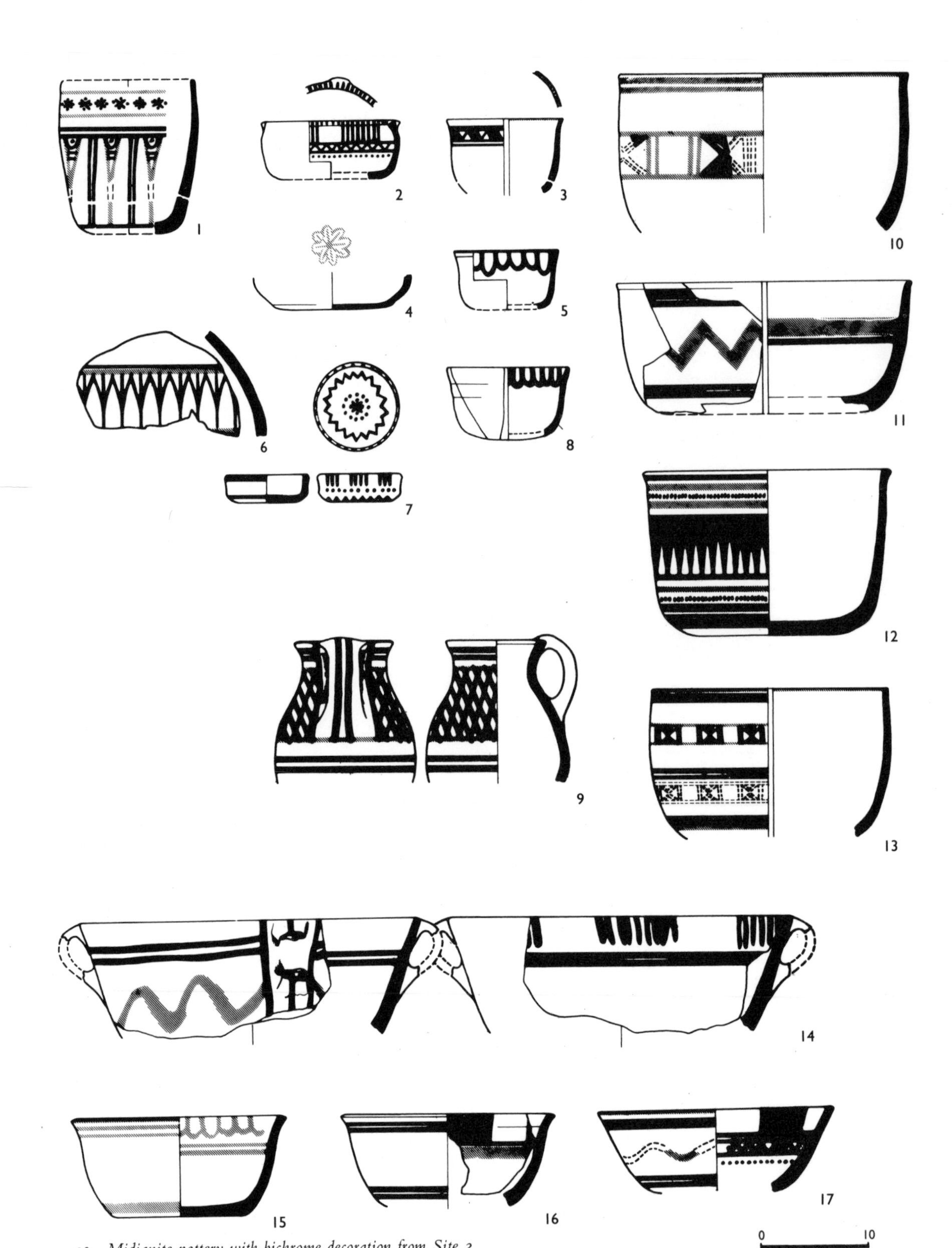

32 Midianite pottery with bichrome decoration from Site 2

thickened bases can also be dated to Late Bronze Age–Early Iron Age tradition. Two pithoi with collared rims are of a highly indicative type very common in the Early Iron Age of Palestine. Other typical types of Late Bronze Age date are jugs with pinched or rounded mouths, a pyxis, and some fragments of deep lamps with small rims and rounded bases.

Only few details were added by the excavations to the repertory of the Negev ware, including some small cups and bowls with inverted rims and rounded bases, some round-based cooking pots and several vessels with fabric impressions not only on their bases but also on their sides. This latter detail clearly shows an additional method of primitive pottery manufacture in Timna. Conclusive evidence for the origin in Timna of at least part of the Negev-type ware found in the excavation, can be seen in the copper slag fragments included as temper in the clay of many of these cooking pots.

Many new pottery types and decorations were added to the repertory of the Midianite pottery, including a number of whole vessels. From these vessels it became possible to learn something about the Midianite method of pottery making: traces of a turning process can be seen on some of the larger bowls but these are different from the lines left behind by the fast potter's wheel. It therefore seems most likely that the vessels were made on a slow revolving base, or even only smoothed by a slow turning process. A thick slip was applied in pink-buff or yellow-brown and well smoothed.

Fig. 32

Plates 48–52, 54

Among the complete vessels found were cooking pots with flat bottoms, straight sides and slightly flaring rims and without decoration, but made of the typical Midianite buff to light red clay, usually slipped and burnished. Additional patterns of decoration add many variations even to the most common combinations of horizontal and vertical lines, wavy and angular zig-zags, dots, crosses and entwined and concentric semi-circles. There are arrangements of metopes and triglyphs with diagonal lines, and alternate criss-cross hatchings or filled triangles. A complex design is applied to some deep cups, consisting of long narrow bichrome triangles with an 'eye' in the upper end, enclosed on three sides by dark brown double lines. A line of crosses and dots in brown, between two red lines, drawn in the upper part of the cup, adds a particular attraction to this design.

The study of the pottery established a Late Bronze Age–Early Iron Age date for Site 2, the lowest possible absolute date being the twelfth century BC. The few datable metal finds, especially the early type toggle-pin and the spear-butt, correspond well with this dating, whilst the two Ramesses II scarabs indicate a thirteenth century BC date. The copper works at Site 2 are, therefore, dated to the thirteenth to twelfth centuries BC.

The study of the sequences of superimposed floors and their relationship to several building and repair phases, as established in most excavated areas of Site 2, together with a minute stratigraphic recording of all sherds found, allows us to draw the conclusion that there is only one archaeological level here. This means that on and in every floor of this level, as well as in the pits, the same Late Bronze Age–Early Iron Age pottery was found.

Buildings and smelting installations were first erected on the red Nubian sandstone covered by a thin layer of red sand. Here, the lowest working-floor was formed with pits penetrating into the rock below. Continued industrial activities led to the formation of several superimposed floors, on each of which remains were found *in situ* and also a number of pits, penetrating through the floors beneath. The number of working-floors varies from area to area but the distance from one floor to the other does not generally exceed 10–20 cm. The deposits on the working-floor consist chiefly of sand, mixed with ash, charcoal and slag particles, bones, industrial waste and domestic refuse. Often a layer of fine wind-borne or water-carried drift sand, seldom reaching a thickness of more than 1–2 cm., indicates seasonal interruption of the smelting activities.

The brown stratum, common to all working-floors, is seldom more than 60 cm. thick, and ends with an uppermost working surface on which were found a number of objects, as well as installations and pits from the final phase of organized industrial activity at the site.

After a prolonged period of intensive, but seasonally interrupted activities, resulting in the series of superimposed, brownish working floors, a violent earthquake struck the site, causing great havoc. Walls collapsed and buried beneath their debris a great many objects and installations. After the destruction no elaborate repairs were made, instead new installations were put up near the demolished, earlier ones. Only minor repairs are evident here and there as, for instance, in storehouse Locus 1021, or in the casting workshop Loci 1037–1038 of Area K. The storerooms, containing water-cisterns or large working installations, were cleared of the debris but otherwise much of it was simply left where it had fallen and new walls built on top or alongside.

There seems to have been a prolonged interruption of work towards the final activities at Site 2. Above the brown stratum of working-floors, a layer of reddish-grey wind- and water-borne drift sand was deposited, with a hard crust on top. Traces of metallurgical activities were found on this uppermost surface, mainly between the protruding tops of drift-sand covered building walls belonging to the brown main stratum. Here and there primitive, mostly circular structures were constructed from stray debris immediately on this uppermost surface, as was the round structure in Area B2–3. Yet the pottery found here, together with stone implements, furnaces and slag, does not differ in any way from the sherds found in the main stratum underneath. There are convincing indications that the people who worked at the site during this final phase of activity, knew about the remains underneath the drift sand layer and made use of some of the still serviceable installations, and also retrieved many of the stone tools left on the uppermost, brown working-floor. It seems that sometime after organized activities at Site 2 had ceased and drift-sand had started to settle on the installations, local workers, who had worked here previously and knew well the details of the site, returned for an additional period of work. These people are assumed to have been the Midianites of the twelfth century BC who were also found to have re-occupied, for a short while, the Egyptian mining temple of Timna and turned it into a Midianite shrine.

IV

High Places and Rock Engravings in the Timna Valley

The three ethnically different groups, which collaborated in the operation of the Early Iron Age copper works, are represented in different places of worship discovered and excavated in the Timna Valley. Some of these are obviously cult sites, for others there seems to be no other plausible explanation, but all need a lot more study for their final appreciation, especially as regards the details of the actual ritual and their ethnic connections.

A small Semitic temple at Area A, Site 2

South-east of the actual industrial area, slightly higher up on the slope and obviously intentionally isolated from the smelting installations and workshops, a tumulus of stone debris attracted the excavators' attention during the first survey of the site. Before the excavation, remains of a small, square stone structure (Structure II) could be seen standing on top of a larger, ruined and sand-covered structure (Structure I). The tumulus was partly excavated in 1964 and cleared to bedrock in 1966.

Plate 113

Structure I, built of local red sandstone blocks of medium size, was a rectangular building, 9 m. long (east-west) and 8 m. wide, with its entrance at the east end. Next to the entrance, inside the building, a low stone bench, possibly an offering bench, was built against the eastern wall. A large, square, flat-topped stone stood in the centre and served, presumably, as an altar. All around it were ashes, a large quantity of broken animal bones, fruit kernels, some pottery representative of the three groups already mentioned and beads. At the west end of the building five large, rough-hewn, stone slabs were standing in line, carefully kept upright by wedges of small stones under their base. These were obviously standing-stones (*mazzeboth*). A libation bowl stood before them, carefully carved from a sandstone block. The bowl stood on a round, bowl-like cavity, carved into the bedrock, and this must have served as the original libation bowl. One semi-circular annex was built against the outer north walls, another next to the entrance. Much wood ash and

Fig. 33

Plate 114

Plate 110

many crushed bones, mainly of goats, were found inside both annexes.

Structure I appeared to have been a small place of worship, of typical Semitic layout, attached to the large smelting Camp 2. It may also have served the workers of camps 14 and 15, located only a short distance across the ridge to the east of Camp 2, who otherwise do not seem to have had their own cult place. It was built after metallurgical activities had gone on for some time in the adjacent smelting camp, because below its north-eastern walls a thin layer of metallurgical waste was found. The other walls, altar and *mazzeboth* stood on a thin, undisturbed, layer of red Nubian sand or directly on the red bedrock. Originally the walls, now about 70 cm. high, must have

33 Site 2, Area A: A small Semitic shrine found, almost completely empty, next to the copper smelting installations. Structure I is the outer perimeter, Structure II the shaded stones area

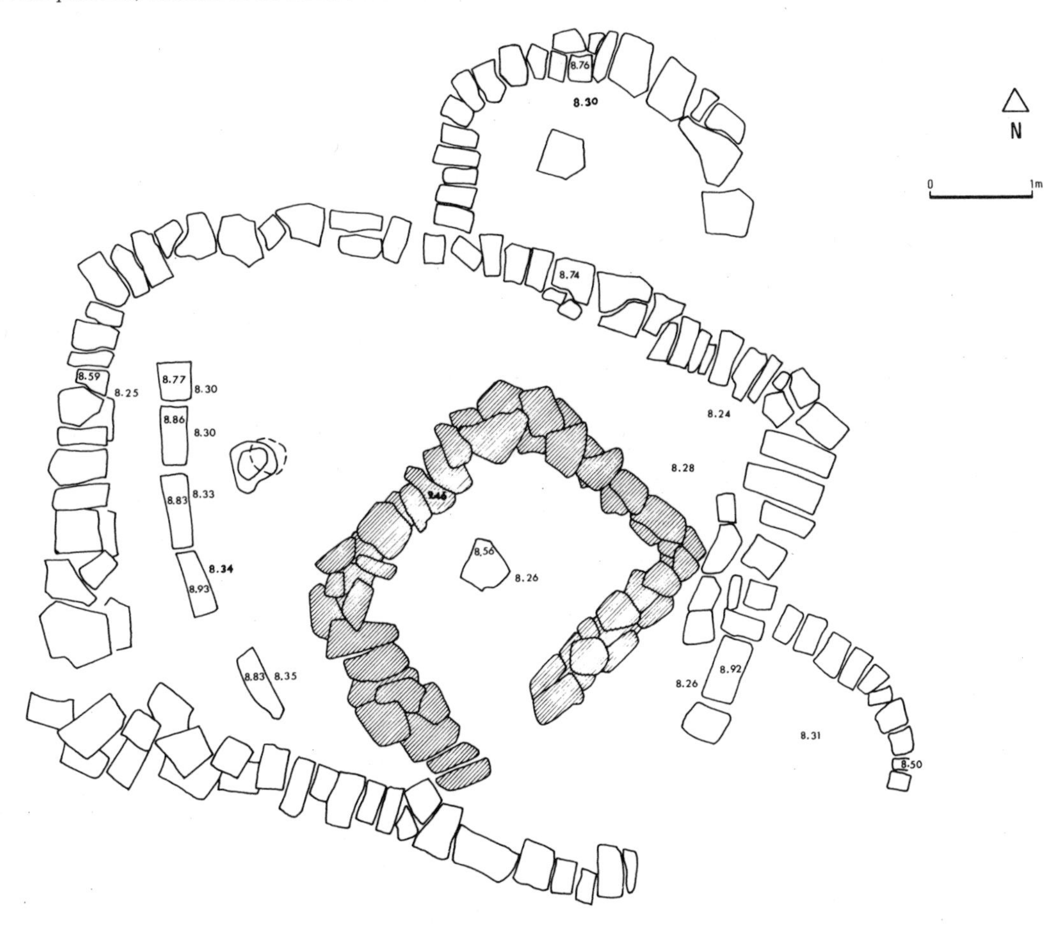

been approximately 1·5 m. high. A mass of their debris was found laying in orderly lines, mainly on the south-west side of all walls. The seemingly sudden collapse of all the walls must have been caused by an earthquake, attested also in other parts of the Timna Valley and at the excavations of Site 2.

The ruins of Structure I became almost completely covered by wind-borne fine sand. By the time this sand-fill had reached the height of the altar, its top still slightly protruding, Structure II was erected on top of the debris, with the altar in its exact centre. This may, of course, be pure chance, but many other indications in the excavation of Camp 2 make it almost certain that the newcomers to Camp 2, who also built Structure II, were familiar with the layout of the camp, its stores and installations, and may have continued using the altar.

Structure II was 3·5 × 2·5 m., its entrance also at the south-east side. Its walls were originally about 80 cm. high. Besides some sherds of the rough non-decorated kind, no finds of any sort were made. Like its predecessor, it slowly filled up with fine wind-blown sand.

A High Place at Area F, Site 2

Fig. 16

Plate 109
Fig. 34

Smelting site 2 is contained in a small valley, enclosed by low, rugged mountains. About 70 m. west of the actual smelting area, a cone-shaped hill, about 20 m. high, ends the chain of hills enclosing the site from the west. The summit of this hill is a flat area of 5 × 8 m. Here, remains of a completely destroyed structure were found, today no more than a group of rough stones, most of the original structure having collapsed onto the steep slope below, forming a large heap of debris. Next to this destroyed structure, an area of about 4 sq. m. was covered by metallurgical waste – small slag pieces, burned stones, ash, fragments of charcoal and a scatter of sherds.

Plate 24

The cone-shaped hill is connected with the main chain of mountains by a narrow saddle of rock, whose maximum width is about 4 m. On this ridge was an oval-shaped tumulus, measuring 4·80 × 3·10 m. and about 50 cm. high. Between this and the slightly higher summit, the ground was covered with a jumble of stones and metallurgical waste, which included burnt earth, fragments of small crucibles with adhering slag, charcoal, clay protectors for bellow nozzles, broken stone crushing tools and lumps of furnace wall lining. From this material, collected both here and in the debris heap below, it is evident that a melting-casting installation was being operated with some of the work taking place higher up, on the summit of the hill itself, next to the destroyed structure.

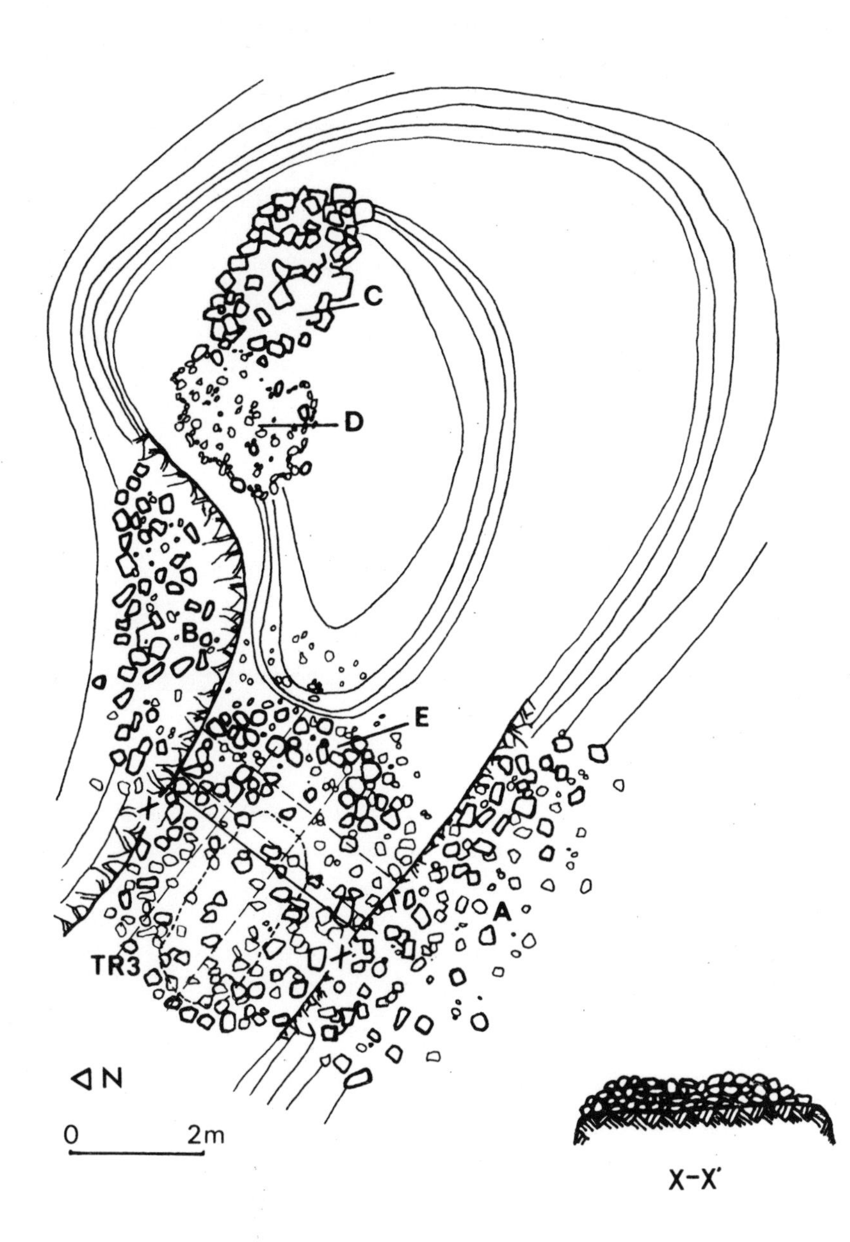

34 Site 2, Area F: Plan of the High Place on top of a hill next to the smelting camp. C debris; D metallurgical waste; E burnt soil, ash, charcoal and slag

Although some metallurgical waste extended over the eastern end of the tumulus, subsequent excavation proved that the tumulus itself was not directly connected with the metallurgical installation. The tumulus was completely cleared to bedrock. Before excavation, the shape of the stone heap gave the impression of a small, rectangular structure, the walls of which had collapsed, mainly onto the slopes below, on both sides of the ridge. A section dug across the tumulus showed a depression in its middle, but no clear wall construction could be discerned. The top layer of the tumulus, about 30 cm. thick, was a mass of small disarranged stones, covering what appeared to be a 'floor' of carefully laid flat stones. This orderly core of the tumulus had a maximum thickness of 22 cm. and there was nothing beneath it, only the solid rock on which it was built. An unusually large quantity of pot sherds was found around and within the tumulus, between the stones of its core, and also in the debris on the slopes on both sides of the ridge. There were very many decorated sherds of Midianite type, fragments of blackened cooking pots and large storage jars and also hand-made, rough cooking pots and bowls of the Negev type. Besides the large quantity of pottery there were other finds, rather unexpected in a copper smelting camp: many beads, made of faience, carnelian, mica schist discs, stone and glass, several very small copper spatulae and needles, perforated Red Sea shells and ostrich eggshells. Above the 'floor' itself, and seemingly put there intentionally, were several goat horns, copper rings, two small iron armlets and many beads. Because of its location, the nature of the structures, and the materials found in Area F, this area may be interpreted as a *bamah*, a High Place, where small copper votive implements, such as we actually found later in the neighbouring Hathor Temple, were cast. Some simple faience beads were also manufactured here. It seems likely that the metallurgical operations, which undoubtedly took place here, were an integral part of the actual ritual and it would appear that the Midianites, the makers of the similar copper votive gifts found in the Hathor Temple, were the worshippers at this site.

A number of perplexing facts should be mentioned in conclusion:

1 Many stone implements were found in Area F. Yet, in contrast to the finds from the excavation and from the surface of the smelting areas below, all the implements found here were broken.
2 Every scrap of pottery was most carefully collected at the site, excavated down to bedrock. This effort was mainly in order to secure some complete Midianite vessels, such as had never

previously been found elsewhere. The final total was 84 large baskets full of pot sherds. Despite this quantity of material it still proved impossible to restore any complete vessels, although it was possible to reconstruct several vessels from a few sherds found in the area.

3 Utilizing pottery mending as a stratigraphic check, particular attention was paid to the sherds found within the 'floor' or the core of the tumulus in relation to those found above it, as well as in the heaps of debris on the slopes. It became evident that some of the sherds found inside the 'floor' belonged to sherds found both above it and also in the debris on the slope. Yet, there can be no doubt that the core was laid from flat stones and had a radically different structure from the stone layer above and the 'walls' on its sides. It is also noteworthy that no metallurgical traces were found in the area of this core.

Site 34, a rock altar

Plate 61

Site 34 (G.R. 14509090), located on top of a flat hill of white sandstone in Nahal Nehushtan, has already been mentioned. Its large mass of finely crushed slag is dispersed over most of the plateau, indicating very intensive metallurgical operations. Only at the extreme north-east end of the hill (at point A), is there no sign of any metallurgy. Here, natural rock steps lead up for about 10 m. to a raised platform, which shows clear marks of tooling. In fact, the top of this projecting rock, 3 × 3 m., was carefully flattened. At its foot, four large and several small shallow cup marks (15–40 cm. in diameter, 10–20 cm. deep) were carved into the natural steps. This platform and the cup marks appear to be the rock altar and libation bowls of a *bamah*, a High Place, which served the smelters of Site 34. Its sacred nature, as a place of worship, may be the explanation for the fact that no metallurgical waste was found anywhere near it and in this respect it is quite different from Area F at Site 2, which was the site of ritual casting. A small, very primitive and seemingly ancient figure of a camel with a human figure standing or sitting on its single hump, found engraved in thin lines on top of the altar, may give some ethnological indications as to the worshippers at this High Place, but further research will be required. It is, nevertheless, obvious that it is not an Egyptian cult place and it may have been used by the Amalekite makers of the Negev-type pottery, found in abundance at Site 34. *Bamah* A, towering high above Nahal Nehushtan, conspicuous from afar, could well have been an inspiring place of worship.

Plate 62

Site 198, a funerary shrine on top of 'King Solomon's Pillars'

On top of the red Nubian sandstone formation, called 'King Solomon's Pillars', it is difficult to move. After climbing upwards for approximately 30 m. on a steep, sandy slope at the south end of the rock formation, the top of the mountain is reached and the going becomes very rugged, with huge boulders and narrow fissures. At G.R. 14579094 the area turns into a labyrinth of narrow passages between straight-sided hillocks of 3–5 m. high. These low hillocks are actually huge blocks of rock, with low, long and narrow empty crevices washed out by wind and water at their bases, many of them closed by small flat stones. In fact, these rock crevices were turned into narrow chambers or cells, most now empty but several are still closed. Scattered human bones indicate that this area, Site 199 on our survey map, must have been a burial place. One complete, large Negev-type cooking pot and a beautifully decorated Midianite jug were found here.

Fig. 35

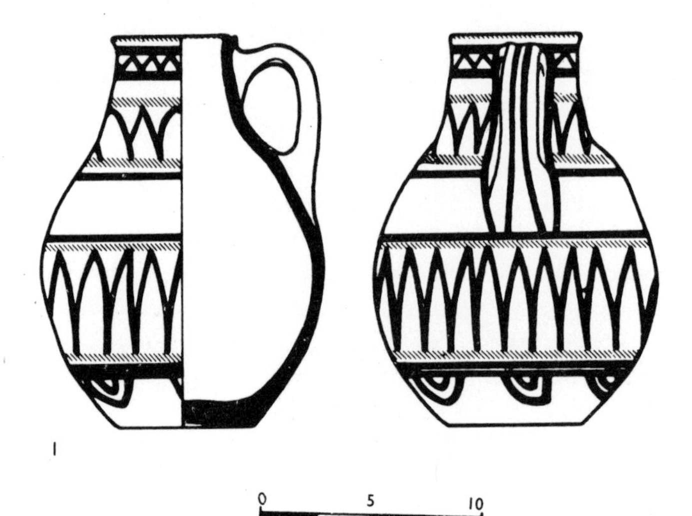

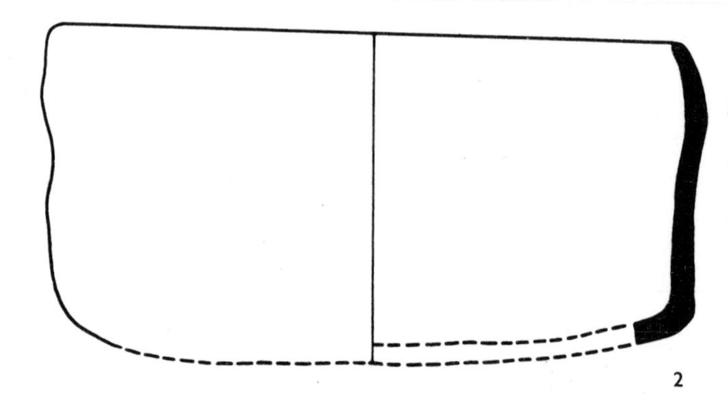

35 Midianite bichrome decorated jug and Negev-type cooking pot found at the burial site above the Timna temple (Site 199)

About 50 m. south of Site 199, a huge block of rock slipped down from above and landed on its side, leaning against the face of the mountain. A triangular, shelter-like space was thus formed between the rock face and the fallen stone, about 4 m. high, 3 m. wide (on the ground level) and about 4 m. deep. Inside this narrow shelter, Site 198, a standing stone, 55 cm. high and 30 cm. wide, with carefully rounded top, was found on a flat, table-like, rock. To keep the standing stone from falling it was held in position by several small wedge stones and it also leant against the mountain face. A flat stone with a very shallow cup mark was lying right at the foot of the 'table'. Inside and in front of the shelter decorated Midianite sherds were found and also several stone tools and a small quantity of slag and charcoal. This site is undoubtedly a small shrine, connected with the burial site on top of 'King Solomon's Pillars', with a *massebah* standing on an 'offering table', and a libation bowl right next to it. Considering the pottery found, it seems safe to conclude that this was a Midianite funerary shrine. The slag fragments, tools and charcoal found could belong to some kind of ritual casting, like those at Area F, Site 2, described above.

Plate 111

Plate 112

Votive rock-drawings in the Timna copper mines

The mining sites 24 and 25 are situated along the northernmost upper reaches of Nahal Timna. Here the Timna cliffs retreat and form a small side valley, flanked on both sides by huge deposits of cupriferous Middle White Nubian sandstone and slopes with very numerous ore-dressing 'plates'. Mine 25 forms a long, rugged triangle, projecting nose-like into the wadi and is conspicuous from afar by its light, almost white tints.

About 50 m. before the southern tip of mine 25, a 10 m. high pile of rubble and large boulders, evidently caused by heavy stonefall from above, lies against the mining wall. Some 5 m. above the head of this pile is a 5 m. wide and 2 m. high rock picture (Engraving 1) engraved on a particularly smooth area of the rock wall.

ENGRAVING I

Plate 59

Fig. 36

Although it is debatable whether all the details of this picture were made by one and the same hand, it is evident that it was done purposefully as one pictorial unit. The figures are deeply incised with a sharp-pointed implement and several groups of small holes are visible within the area of the picture, mainly on the left upper corner. No meaning could be attributed to them and it seems doubtful whether they are related to the original rock engraving. The picture was conceived as three long rows of figures of which the upper and lower rows are well preserved. The middle row has suffered some

Plate 60

0 1m

36 Rock-engraving 1, found on the wall of mine 25

damage and several of the figures are, therefore, difficult to see. A solitary, strange, 'human' figure is carved above the left end of the upper row, its hands raised, showing four widespread fingers only. A curious object, drawn as a straight line ending in an oval sling or handle, projects sideways from his hip, and could represent a weapon. Similar strange figures have been found previously at other Iron Age sites in the Arabah and also at Site 25, Engraving 2, and it is suggested that it may be a magic representation, or some 'higher spirit'. Its isolated position above all other figures should also be noted.

All three rows of figures consist mainly of two types of animals: the ibex (*capra ibex*) and the ostrich (*struthio camelus*), but there are also some gazelles (*gazella sp.*). Several human figures probably representing hunters appear within the rows of animals, some holding a lasso or a shield in their left hand. One ibex is drawn lying on its back. In the centre of the uppermost row there is a curious sign, a square with two short legs. The same, unidentified sign was found at Site 251 in the Arabah. At the right end of the lowest row there is a crude representation of a chariot which, curiously enough, seems to have only an axle with no chariot floor and the draught animals are not really connected to the four-spoked wheels. The 'chariot' is drawn by two ibexes with long, drawn-back horns, harnessed together at their heads by a rather heavy looking cross-beam. Presumably the

Fig. 37

originator of this chariot engraving never saw a real chariot, but copied his from the memory of another nearby rock-engraving, representing several chariots (Engraving 2).

Before turning to this second rock-engraving, a few relevant details of Site 25 need to be described. The rock wall to the left of Engraving 1 shows clear evidence of tectonic changes, presumably caused by earthquakes. At its foot are piled large boulders which have obviously broken away from above, leaving a very rough wall with a much lighter patina. The same lighter patina also continues on the lower part of the rock wall below Engraving 1, which, besides the rubble underneath, can provide evidence that here also some stone masses must have fallen down. Next to Engraving 1, remains of an earlier tubular cistern were found, with a vertical row of footsteps leading up to about 2 m. below the level of the engraving. This indicates the original height of the top of the cistern. Further to the left of the engraving, a ledge in the rock-wall continues for about 10 m., ending at the entrance to a small cave, approximately 8 m. above the present surface. Here another broken cistern was found, its upper rim exactly at the height of the ledge in front of the cave. Inside the cave was a fireplace and some pottery. There can be little doubt that, at the time the rock engraving was actually made, there must have been a wide ledge or even a solid slope running all along the wall and the artist must have stood on it, next to the cistern's upper rim.

ENGRAVING 2
Plate 59

Continuing westwards for about 100 m. along the cliffs, a narrow canyon opens into the mountain side. The canyon is about 40 m. deep, 20 m. high and 3–5 m. wide. On its left side a 9 m. long and

37 Unidentified inscription at Site 251 in the Arabah

0 30

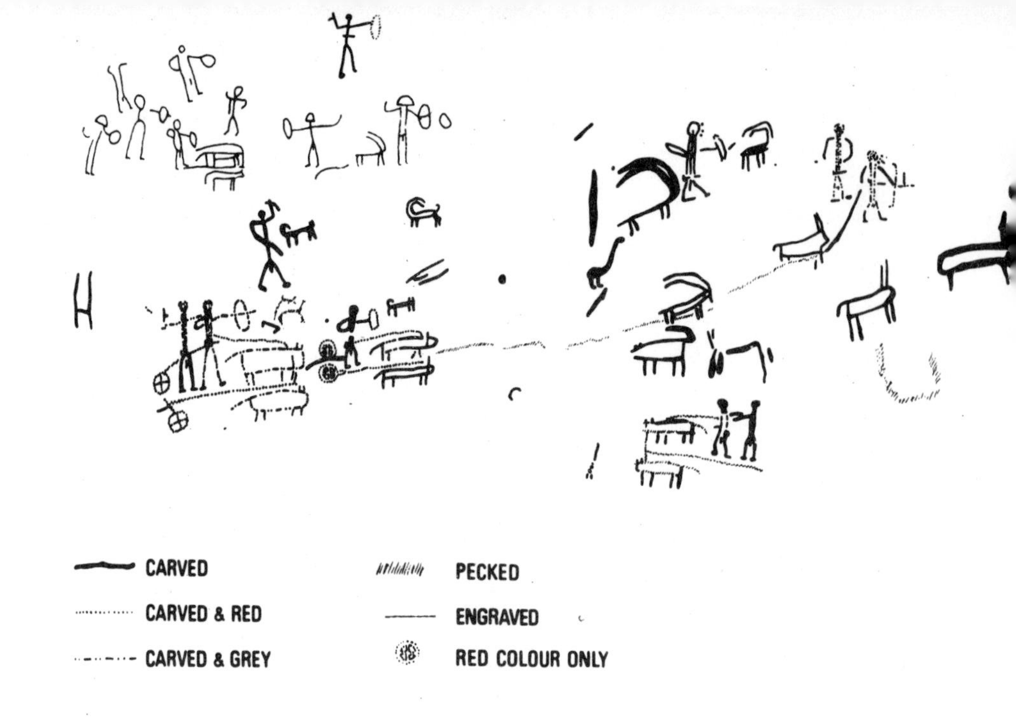

Fig. 38

1·5 m. high rock-engraving (Engraving 2) was cut into a smooth stone frieze, sheltered from the sometimes torrential rain of this area by overhanging rocks.

Engraving 2 is incised into the rock with a sharp point and many of its lines were filled in with red and white colour. Already from the point of view of its engraving-colour technique this picture is most unusual. Nothing like it has ever been reported in the Near East, although the technology of engraving and subsequent colouring is well-known from Egyptian wall-paintings. The details of this picture convey a great deal of information, although much of it remains enigmatic, mainly due to lack of comparative material. The basic theme of Engraving 2 is an arrangement of four-spoked, manned chariots. The chariots are drawn having rear-positioned wheels, but without a side-screen. The draught animals, which seem to be oxen, are harnessed to the front of the pole and appear to wear some kind of yoke. One or two armed men are standing on the floor of each chariot. Some of the charioteers carry a round-topped or circular shield or a bow, whilst almost all the occupants hold a typical Egyptian New Kingdom battle-axe in their raised hand and have a short, hilted dagger. To free the hands of the charioteer, the reins are tied around his waist, a fashion well-known from New Kingdom wall-paintings and also from the Ugarit gold patera dating to the late fifteenth or early fourteenth centuries BC. All occupants of the chariots and the round-headed, battle-axe-carrying

Plate 55

38 Rock-engraving 2, a processional array with apparent magic significance on the walls of mine 25

men standing before them, wear a loin cloth, folded into a pointed apron in front. The artist carefully drew the details of this fashion typical of the Egyptian soldier of the New Kingdom, as can be seen, for instance, on the reliefs of Queen Hatshepsut's temple at Deir el-Bahri.

In the upper centre of the panel we meet men of a different type, their bodies drawn as two long parallel lines to indicate their larger size. These men seem to be wearing some kind of helmet, carry long, hilted, straight-bladed swords, and wear tasseled kilts. This part of the engraving, contrary to the chariot array, is populated by various animals: we have here ibexes, some ostriches, dogs and several representations of the straight-horned oryx (*oryx leucoryx*). The latter has so far not been found in any rock drawings in the area and a small number of this desert antelope are known still to exist in South Arabia. The dogs seem to chase the animals and one of the hunters has just sent an arrow after an ibex. In the lowest part of the picture, next to the third chariot from the right, appears a horse with a rider (?) on its back. The group of barely discernible scratchings in the upper left corner of the engraving, badly copying some of the figures of the original drawing, is an addition, probably later, made in red and by rubbing with a blunt implement. The solitary figure on the extreme right is incised on a different plane of the rock, and may have been added here by the people that engraved the first rock-drawing. A later hand tried to 'improve' on the general layout

Plate 56

Plate 56

of the engraving and pecked a kind of a sloping surface through part of the picture; they also added an unidentified, pecked object near the middle of the picture.

Essentially, Engraving 2 seems to contain two different main themes, each with its own and differing 'actors'. The ox-drawn chariot groups, manned by battle-axe brandishing charioteers, could not possibly reflect any real local event, neither fighting nor hunting; they seem to represent a processional array with some cult or magic significance. Right underneath Engraving 2 were found several large bowls made of very soft white sandstone, broken by rocks that fell from above. These bowls, the like of which were not found anywhere else in the mining areas or in the smelting camps of Timna, could not have served any metallurgical purpose because of their extreme softness. Yet similar bowls or basins were found in the Egyptian mining temple at Timna, apparently used there in a ritual context. A quantity of sherds and bowls found in the canyon near the engraving belong, like the Timna Temple described below, to the XIX to XXth Dynasties.

Plate 54

The second theme of Engraving 2, concentrated in its centre, seems to be a hunt, with men and dogs chasing ibexes, ostriches and antelopes. It is obvious that the artists intended to differentiate between the Egyptian (?) charioteers and the tall, tassel-kilted hunters. These hunters may be identified with the Shosu of the Egyptian sources, perhaps here the Midianites, inhabitants of southern Transjordan and the Hedjaz. A date in the XIX to XXth Dynasties for Engraving 2 fits well with the date proposed for Engraving 1. Besides, pottery of this period found in the rubble underneath Engraving 1 should, by all the evidence, be contemporary with the rock-carved cisterns found next to it. These cisterns are known to belong to the Egyptian mining enterprises of Timna, dated by the Timna Temple to this period, the end of the fourteenth to the middle of the twelfth centuries BC. Whether the human figures on Engraving 1 represent people ethnically different from the two human types of Engraving 2 and therefore a third ethnic factor, perhaps the Amalekites from the Central Negev mountains, known also to have worked in the Timna mines at the same time, is difficult to ascertain. The strange copy of the chariot on Engraving 1 seems to indicate that Engraving 2 must have been either earlier than or at least contemporary with Engraving 1, and the date proposed for both rock-engravings seems to be well established and their votive character plausible.

V

The Hathor Temple at Timna

Location and method of excavation

During the detailed survey of the Timna Valley in 1966, attention was paid to the vague outlines of what appeared to be the foundations or ruined walls of a small structure built against 'King Solomon's Pillars' in Nahal Nehushtan. The 'Pillars', mentioned already above in connection with the Chalcolithic mining operations in the Arabah, are huge, picturesque, eroded formations at the south-western end of the Timna massif, almost in the centre of the ancient mining and smelting area of Timna. The site, numbered 200 on the survey map (G.R. 14579090), was a low mound, measuring 15×15 m. and only approximately 1·5 m. in height, leaning against one of the 'Pillars'. Some white sandstone structures protruded slightly above the present surface and here and there some sherds, mainly Midianite, and a small copper arrowhead were found among the piles of rubbish left at the site by the many tourists visiting the 'Pillars'. The fact that a building was erected at all at this particular spot, and that it was built of white sandstone (which had to be carried here from quite a distance) instead of the available local red sandstone, was very peculiar. Furthermore, three niches cut into the rock of the 'Pillar' immediately above the mound, in the shade of a huge rock-ledge overhanging a great part of the mound, added further interest to this unusual site.

Plates 65, I

Plate 66

Excavation of the site was begun in March 1969. First, a considerable quantity of recent refuse was removed and a one-by-one metre grid laid over an area of 13×29 m. This was excavated to bed-rock with the exception of the building remains themselves, which were left standing, and part of the 'white floor' in Loci 109–110, which was also left *in situ*. A cross of two baulks was carefully kept, right to the end of the excavation, and drawn as stratigraphic sections. One ran NW–SE, from the large niche in the face of the 'Pillar' to the lowest end of the slope below the structural remains; the second ran NE–SW, parallel to the face of the 'Pillar' and right across the building and the slopes on both its sides. Each metre square was excavated as an independent area and the finds recorded

Figs 39, 40
Plate 67

Plate 69

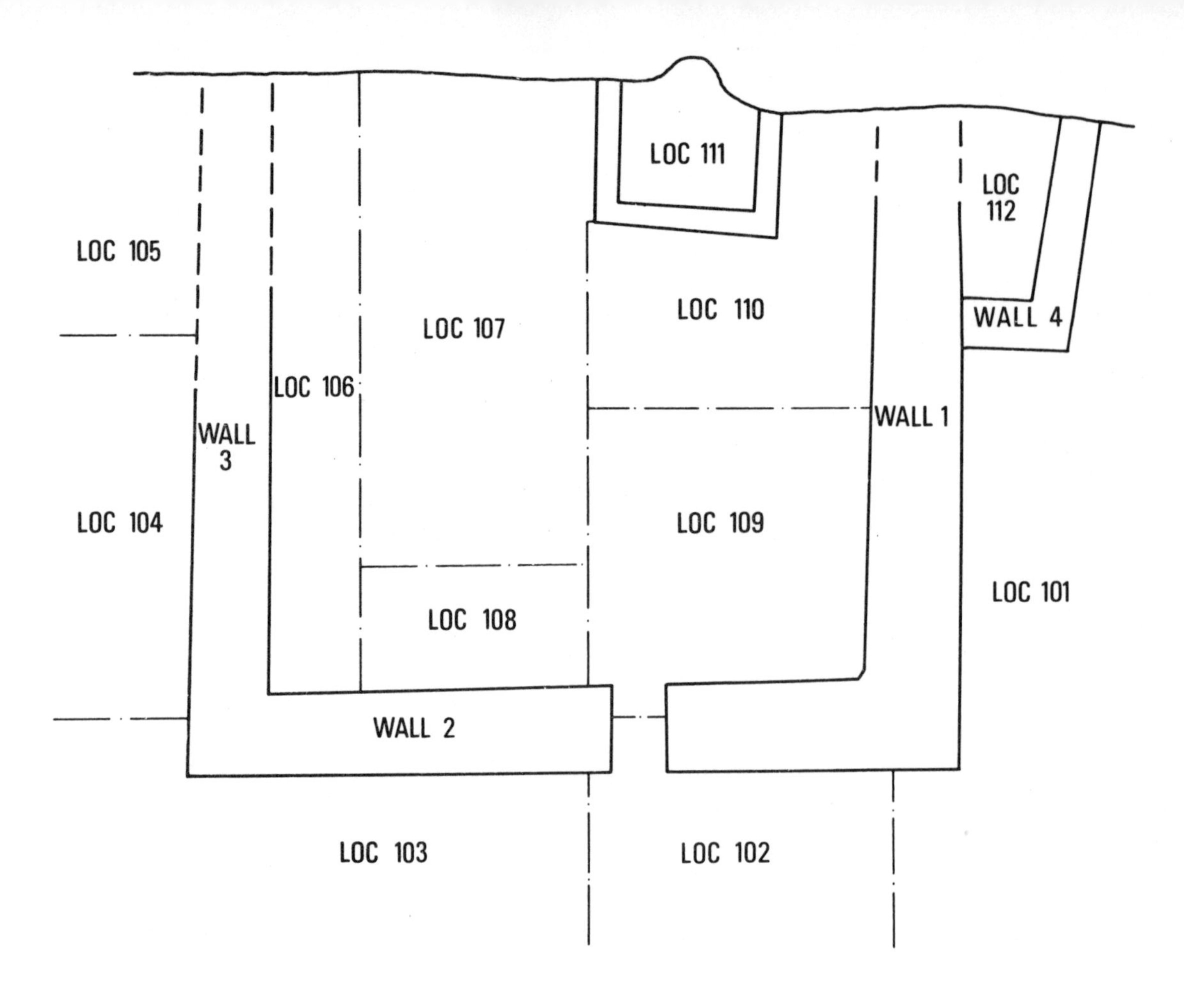

39, 40 Site 200, The Temple: Plan of excavated loci (Locus 111 is the temple naos), and a simplified section of the temple, showing its location in relation to the rock wall and its major occupation layers

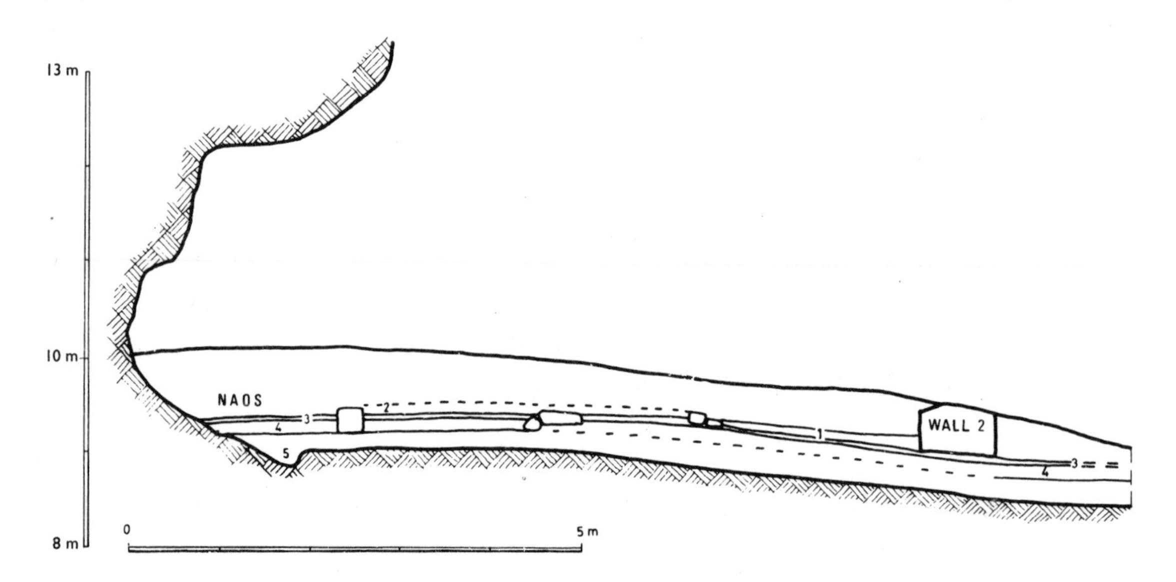

accordingly. The finds on floors, or floor-like surfaces, were usually left there until the whole area was cleared to the same surface in order to present, even during the excavation, a concrete picture of finds relative to structural remains on the same floor. Although the finds were individually recorded in their appropriate squares, a separate record was also kept of these interrelated data.

The excavation of the small mound was originally planned as a trial dig and scheduled to take about two weeks. Yet, right from the first day unusually numerous, uncommon, small finds were made, especially on the northern slope of the mound, necessitating that all excavated sand be sifted through two sets of sieves. The result was a prolonged season of more than two months, and almost 10,000 small finds. It often took several days to clear down to a depth of about 50 cm. because of the innumerable finds, including hundreds of tiny beads, very fragile faience, glass and wooden objects, not to speak of masses of decayed cloth. The main reason for the prolongation of the excavations was, of course, the nature of the site itself. On the fourth day of work a rich hoard of metal objects came to light, including copper and iron jewellery, mixed with a large quantity of specially chosen ore nodules, and Egyptian faience objects, some of the latter bearing hieroglyphic inscriptions. A figure of a non-Egyptian male god (?) was found in this hoard, as well as some Midianite sherds. The next day a small copper snake was found in Locus 111 and, immediately afterwards, the excavating team were amazed to find an exquisite faience face of Hathor, the Egyptian goddess, about 20 cm. below the present surface in Locus 110. This find made it obvious that the site was an unusual place of worship; in fact, the several representations of Hathor already discovered during the first days of the excavation made it evident that here an Egyptian temple, dedicated to Hathor, was being excavated.

Plate 95

Plate XVIII

Plates XIX, XX

Plates 80, X

The stratigraphy of the Temple site

The actual habitation layer of the mound consists of sand, structural remains, floors and floor-like surfaces, destruction debris and many unrelated building stones – all these in a layer of less than one metre deep. If we add here that within this layer remains were found which date from the fourth millennium BC to the Roman period, including destruction and secondary uses of structures and floors, the complicated stratigraphy of the site becomes evident. Nevertheless, it was found possible to relate many of the finds to certain chronological phases of the site and, in many cases, to certain architectural features as well.

Fig. 40 Five distinct habitation-phases could be distinguished in the excavation. Starting from the top of the mound the following five archaeological strata found were:

Fig. 41 *Stratum I*, the latest phase, a secondary use of the temple site which took place in the Roman period, approximately the first century AD.

Stratum II, representing a period of great upheaval in the history of the site, of destruction, a devastating earthquake and a short, final revival of the Egyptian temple as a Semitic, probably Midianite shrine. This final phase is dated not later than the middle of the twelfth century BC.

Strata III and IV, the main phases of the original construction, later destruction and subsequent rebuilding of the Hathor Temple, dated to the XIX–XXth Dynasties of Egypt, from the end of the fourteenth to the middle of the twelfth centuries BC with a break of perhaps a

41 *Site 200: Plan of the Timna temple in its Midianite phase (Nabataean remains are not shown)*

generation in its occupation. It is suggested that this lacuna coincides with the first destruction of the Hathor Temple, during or shortly after the reign of Sethos II, at the end of the thirteenth century BC, although its accurate archaeological dating was not possible.
Stratum V, the earliest occupation of the site, dated by pottery and flint implements to the Chalcolithic period.

The story of the Hathor Temple

ARCHITECTURE

The archaeological stratification, architectural remains and habitation-surfaces, combined with the analyses of the numerous finds from all strata of the mound, and compared with the data gleaned from the many inscribed objects found in the excavation, result in a fairly clear picture of the life story of the Timna Temple. It falls into four phases:

Plate 73

(*a*) At the beginning, there were several shallow rock-cut pits and a few fireplaces under the rock-shelter formed by the huge overhanging walls of 'King Solomon's Pillars'. Some structures seem to have existed here but only meagre evidence of them was found under wall 3 in Locus 105, giving no idea as to their possible function. Near the fireplaces, but mainly in the bottom levels resting on the sandstone bed, a number of Chalcolithic flint tools, some rope-decorated sherds and fragments of Chalcolithic hole-mouth jars were found. The flint implements were mainly scrapers, but included an awl, a borer, blades, cores, and hammerstones. A large number of flakes and other flint waste testify to the manufacture of flint implements at the site during its Chalcolithic occupation. Whether this site was already a Chalcolithic place of worship or whether it served merely as a convenient camping site in the shade of the overhanging rock, could not be established by the excavation. Not enough of the earliest phase could be unearthed without destroying the temple structure and this problem remains, therefore, undecided. However, it seems that already in this earliest period of copper smelting in the Arabah, the Chalcolithic miners of Timna dedicated this unusual location to some special purpose. Nowhere else in Timna were similar remains and rock-cut pits found, although other camping sites with Chalcolithic flint tools and pottery were located along the slopes of Har Timna and also along the bottom of 'King Solomon's Pillars'. Perhaps it is not a matter of fortuitous coincidence that inside the naos a shallow pit with Chalcolithic remains was found below the earliest Egyptian surface.

Fig. 42

(*b*) More than 2000 years later, during the reign of Sethos I (1318–1304 BC) of the XIXth Dynasty, an Egyptian temple was erected on top

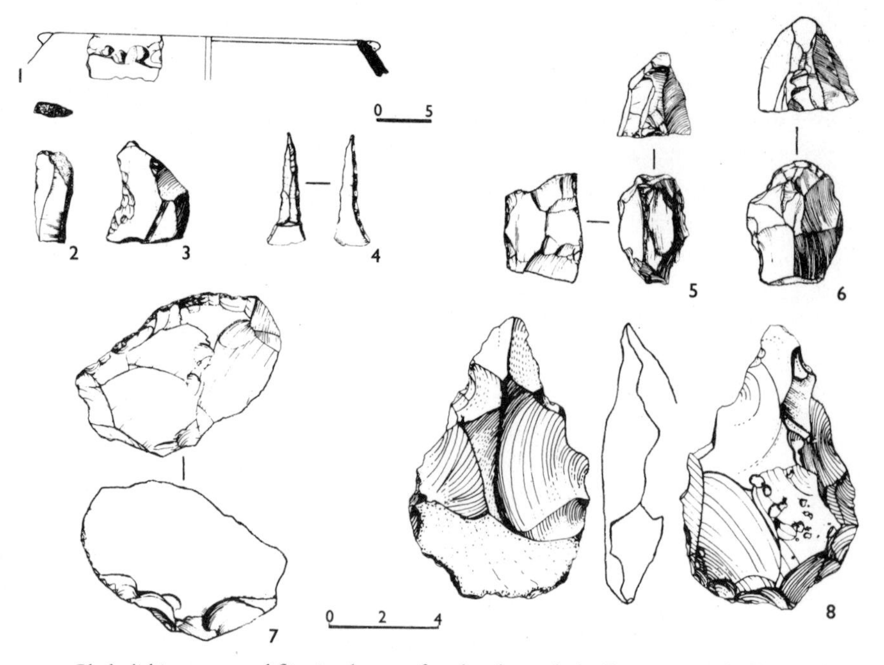

42 Chalcolithic pottery and flint implements found underneath the Egyptian temple (Site 200)

of the Chalcolithic remains. An open court, measuring 9×7 m., and mainly of red Nubian sandstone with some limestone boulders was built against the face of the 'Pillar'. It was a dry built, rough, but very solid structure. There is no evidence for the location of the entrance to the court at this initial stage of the temple, but it may be assumed that it was more or less at the same place as in the second temple phase, *i.e.* in the eastern wall, opposite the naos. The naos, a small shrine, 2·70×1·70 m., of white sandstone, leant against the face of the 'Pillar', with a large, almost man-high niche in its centre. Very little masonry of the original naos remained *in situ*, but two well-dressed square bases and several additional basic parts of the original structure helped in establishing some of its details. Two square pillars, bearing sculptured representations of the head of Hathor, were found in the excavation and seem to have stood on the bases. One end of a large stone architrave, parts of which were found in the excavation, had rested on the Hathor pillars, whilst the other end rested in the two niches cut for this purpose into the rock-face.

Plate 68

Plate 78

Plate 74

A considerable number of well-dressed and also several finely ornamented architectural elements were found on the site, mainly in Loci 106–107 and 110, and also dispersed over a wide area around the actual temple site, testifying to the unusually high aesthetic quality of the original naos building. Several square pillars bearing traces of hieroglyphic inscriptions were found lying around the naos, others

Plate 76

were found built into the second, later temple structure but their hieroglyphs were almost entirely erased. A number of finely tooled Egyptian incense-altars were found in secondary use in Locus 106. Here also a flat, rectangular offering table of white sandstone, 72 × 56 cm., with a groove around its edges, was found on the floor. Apart from walls 1 and 3 and some parts of the naos, no architectural details of the original first temple structure survived what must have been a thoroughly wanton destruction; and there was no archaeological evidence to identify the destroyer of this first temple.

Plate 72

(*c*) A new temple was built on top of the ruins of the original temple structure, utilizing some of its architectural elements. A new floor, consisting mainly of crushed white debris from the original temple structure was laid. Next to the naos, and especially in Loci 109–110, it was found to be a solid mass of white crushed stones, almost 15 cm. thick, which became gradually thinner further east and on the slope outside the temple court. This white floor greatly helped to identify the constructional changes and repairs of the second temple, as it was also found underneath some of the walls and structures. Unfortunately, it was of little help in dating the destruction of the first temple as the effects of the destruction went right down to bedrock and few traces of the original floor were found. Although on this early floor, beneath the later white floor, many finds were made, no inscribed and therefore reliably dated objects could be clearly related to it. The white floor lying on the very soft sand of the original court surface seems in itself to have caused a considerable mix-up of any existing stratigraphic evidence.

Fig. 41

Plate 69

Court walls 1 and 3 were repaired and lengthened from 7 to 9 m. and wall 2 was rebuilt. White sandstone, including well-dressed stones from the first naos structure and even some parts of broken Egyptian altars, was now re-used for the repair and enlargement of the temple walls. The entrance was made opposite the naos in wall 2. The destroyed naos itself was rebuilt, and the pillars and ornamented lintels re-used as building material with plaster finish. A stone platform of large, flat, well-dressed stones in front of the naos, measuring about 3 × 3 m., indicates a vestibule or pronaos. This platform was also built of re-used stones, some bearing very fine toolmarks of comb-shaped metal chisels or picks.

Plates 68, 69

The stratigraphy of the naos demanded painstaking clearance of the inside, down to the bed-rock and also several trial trenches into the adjoining pronaos. Inside the naos the existence of three hard superimposed surfaces, with finds lying on them, could be clearly established. The lowest phase is the bed-rock with a rock-cut pit and

Fig. 40

43 Recently discovered rock-engraving of Ramesses III making an offering to the goddess Hathor. A damaged inscription runs beneath the feet of the two figures

some Chalcolithic pottery. On top of a layer of fill a hard surface was found upon which the two pillar-bases of the first temple phase were standing; some pottery was found there besides other traces of habitation. Eleven centimetres higher, another hard surface with pottery and other finds seems to represent the floor of the second temple phase. On top of this there was fill and debris, which had fallen in from the sides only, including plaster fragments from the second Egyptian temple and also some Roman pottery (in the upper level of the fill).

Outside the naos a trial trench was dug close to its front and a similar stratigraphic picture appeared with two superimposed surfaces at almost the same levels as the floors inside the naos. The lower floor I was at the level of the pillar-bases, with pottery, ash and fragments of Egyptian glass. Above this habitation floor several centimetres of fine sand fill indicated a lapse in the use of the temple. On top of this the thick and solid mass of crushed white sandstone fragments formed floor II. There were finds also within the white floor and these must therefore belong to the destroyed first temple. On the second floor in the trench in front of the naos a cartouche bearing the name of Ramesses III was found, as well as Egyptian glass fragments and other finds. It therefore seems that the first temple was built by Sethos I, destroyed or abandoned during the reign of Sethos II, and the second temple was reconstructed by Ramesses III. However, it should be remembered that there is no definitive archaeological evidence for this although it does appear to be a logical solution which fits the historical picture of this period and the meagre stratigraphical evidence.

Although not much progress has been made so far in the actual reconstruction of the naos, all evidence points to a kiosk type structure, with perhaps a vestibule in front serving as entrance passage. There was no need for any roof to the shrine because the overhanging rock of the 'Pillar' served as excellent protection against any rain. Judging from the number of flat, well-dressed white sandstone slabs found it seems that the shrine was encased by stone screens and possibly also the face of the rock on either side of the niche. The central niche in the naos was found empty but perhaps one of the stone figures of Hathor, found thrown into the sand near the naos, originally stood in it. A small sphinx, the upper part of which was found next to the naos, may also have stood inside the niche. Both stone sculptures were made of local white sandstone. The face of the sphinx, together with the cartouche on both its sides, were unfortunately wantonly erased, but it bears a resemblance to Ramesses II.

Plates 77, 81

Plate 79

65 The Hathor Temple was found under the rock overhang at the foot of 'King Solomon's Pillars' (to the left of the Land-Rover). The photograph shows the site several years before it was recognized and excavated. It was the series of rock-cut niches and the scatter of white sandstone and potsherds that first drew attention to the site, leading to its excavation. Recently, a rock-drawing of Ramesses III and Hathor has been found some 30m. above the temple (*Fig. 43*)

66, 67 The Temple site showing (*above*) the rock-cut niches and some ancient surface debris. After removing the top layer of modern rubbish the basic outline of the building became visible (*below*)

68, 69 *Above*, the central shrine (naos) of the temple, built of white sandstone with the three niches in the background. *Below*, the final Midianite phase of the Hathor Temple, seen from the east

70, 71 *Above*, Midianite offering-bench built against the temple wall with, in the foreground, one of the stone-lined post-holes for the tented shrine (*Fig. 44*). *Below*, remains of a Nabataean melting installation in the temple courtyard

72, 73 *Above*, row of *mazzeboth* in the Midianite shrine, incorporating re-used Egyptian incense altars. *Below*, Chalcolithic remains in a pit below the foundation wall of the first phase of the Egyptian naos

74-76 Architectural elements from the Hathor Temple. *Above*, a side view of a white sandstone architrave decorated in typical Egyptian style. *Above right*, a base with part of a small sandstone headless sphinx. *Right*, the damaged remains of a Ramesside cartouche on a large decorated building stone

77, 78 Many parts of the beautiful central shrine of the temple were found scattered around its ruins. They included several representations of the goddess Hathor. *Above*, a small head of Hathor (Plate 81), made of local sandstone, seen *in situ* to the west of the ruined naos. *Right*, a square pillar with a face of Hathor on each side was re-used as a standing stone by the Midianites, who deliberately defaced the goddess's features (*cf.* Plate 72)

79-81 *Above*, the head of a sandstone sphinx, possibly representing Ramesses II. *Below*, small faience votive mask of Hathor (Plate X), and Hathor in white sandstone from the naos (Plate 77)

82, 83 Part of a faience sistrum with, *left*, the cartouche of Sethos I (*Fig. 48*, 2; approx. natural size)

84, 85 *Left*, faience ringstand with an inscription and double cartouche of Ramesses III (*Fig. 49*, 2; *cf. Fig. 62*). *Right*, fragment of a faience bowl inscribed for Merneptah (*Fig. 49*, 1)

86-89 Faience objects representing Hathor: *above*, *left*, her name and title 'Lady of the Turquoise' on a sistrum handle; *right*, her name on a bracelet fragment (*Fig. 51*, 4 and 1). *Below*, Hathor faces from sistra with her typical cow's ears

90 Part of a ceremonial faience wand

91 Three fragments of a faience bowl with a sketch of a leopard looking back towards an unidentified inscription

92 Bone figurine of the Egyptian goddess Bastet

93 Lower half of an Egyptian ushabti figure inscribed with part of the Sixth Chapter of the Book of the Dead

94 Small faience figurine of a cat. The animal was connected with the Hathor cult and was also sacred to Bastet (Plate 92). Numerous fragmentary representations of felines were found in the temple area

95 Immediately outside and against the east wall of the temple a rich hoard was found. It included copper and iron jewellery, objects of faience, some of them with hieroglyphic inscriptions, and the copper figure of a male idol (Plates XVIII, 102–104)

96 A finely made wooden comb, probably an import from Egypt, found in the temple courtyard

97 Midianite copper figurine of a sheep, before cleaning. It was found to be cast and finely polished, with a hole through its neck for suspension, possibly as an amulet

98 A string of faience beads, many thousands of which were found in the excavations

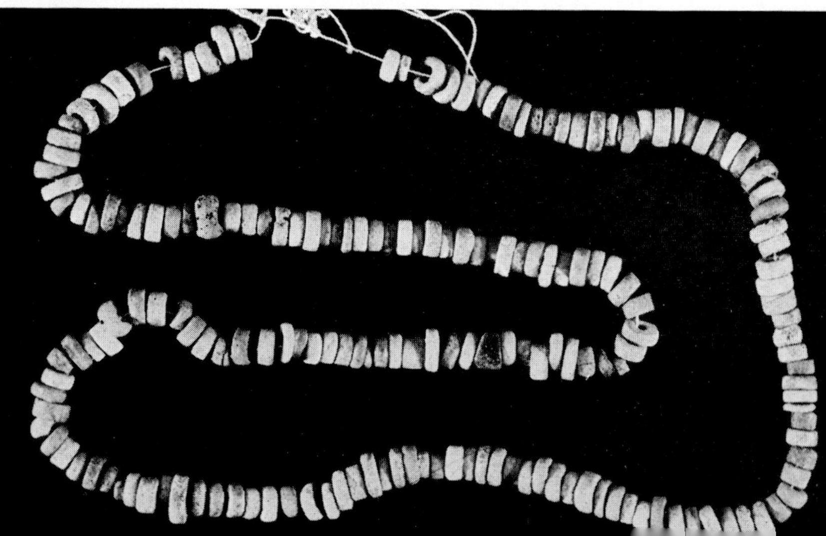

99, 100 Primitive copper castings were brought as votive gifts to the temple. It is possible that they were run-offs from the furnaces and attracted the smelters' attention by virtue of their curious shapes

101 Sherd of Midianite pottery with a schematic drawing of a human figure (*Fig. 47, 4*). It has wide-spread fingers and a strange head-dress

There is a clear similarity between the Hathor temple of Timna and the Egyptian temple excavated at Serabit el-Khadem by Petrie. In both the court wall surrounding the temple was roughly built with undressed stones. Since the excavation considerable thought has been given to the Egyptian origin of these rough, and not quite symmetric walls. However, all the stratigraphic evidence points to them belonging to the initial phase of the temple, with some changes in its second Egyptian phase.

As in the 'Hathor Hanifeyeh' at Serabit, large stone basins, perhaps libation bowls, were found standing in the court of the temple though, with the exception of the basin in Locus 107 next to the naos, one cannot be sure whether they were still *in situ*. In fact, the basin in Locus 110 was clearly in secondary use, probably as late as the Roman period. The basin in Locus 106 might be in its original position, but could also have been moved to its present position during one of the later phases of the temple, perhaps by the Midianites. Its interior was found lined with a layer of mortar.

Fig. 60
Plate 72

Large libation bowls or basins seem to have been connected with the Hathor ritual as many such basins, bearing the actual image of Hathor, have been found at Egyptian temple sites. The fact that most basins in the Timna Temple were carved out of very soft white sandstone and could not have been used for any industrial purpose, should also be considered in this connection. It may be recalled that similar white sandstone basins of ritual significance were also found near the rock drawings at Site 25 in Timna.

Plate 69

Plate 57

The basin in Locus 107 was found broken into several parts, with a thick layer of plaster around its circumference. This use of plaster accords well with the second phase of the adjoining naos, characterized by the use of white and red plaster. As much heavy debris from the collapsed naos was found lying on top of the broken basin it might have been the direct cause of its breakage, and more evidence for the assumption that the second phase of the Egyptian temple was destroyed by an earthquake, clear traces of which have also been found at Site 2 and at other sites (95, 25, etc.).

(*d*) The earthquake which interrupted work at Site 2 and destroyed its shrine (Area A) must also have struck heavily at the Temple site. Although all the court walls seemed to have received a considerable shock at the time, which caused some collapse and the walls to get out of line, it was evident that the west side of the temple was worst hit. The naos collapsed and most of its debris fell to the west on to the floor of Locus 107 and was immediately covered over by stonefall from above. It is probably due to this fact that most of the beautifully

Plates 75, 76

decorated building stones, sculptures and many other fine finds which were covered by the earthquake debris were preserved and not effected by the later intrusions into the temple court, as was the case at the eastern half of the temple and in the naos itself. It seems that after the earthquake the temple was temporarily abandoned and some fine drift sand began to settle on its remains, thus indicating a break in the continuity of the site.

Fig. 41
Plate 70

When the temple worship was renewed, which according to the archaeological evidence must have been a short while after its actual destruction and abandonment, the temple structure underwent considerable and most significant changes. A stone bench of two courses of flat stones was built against the inside of wall 2 on both sides of the entrance. This bench was a rather flimsy structure and could not have served any domestic purpose, yet it was well suited as an offering bench. Wall 1 was breached near the rock-face of the 'Pillar' as an opening into an additional structure, Locus 112, built at this phase against the outside of wall 1. Wall 4 was stratigraphically well above the foundation level of the adjoining wall 1 and was even standing somewhat above the white stone destruction level, found also under part of wall 4. It is assumed that Locus 112 served as the priests' chamber.

Plate 72

The most radical change in the ritual character of the temple was the erection of a row of standing stones as obvious *mazzeboth*, along the inside of wall 3, and consisting of different architectural elements, almost all in secondary use. At this phase in the life story of the temple, wall 3 was no more than a mass of red sandstone debris, fallen mainly outwards and forming a slope against the remaining few courses of the original wall. This was never much repaired but a low bench was built against part of it inside the court and continued as a row of single flat white sandstones right up to the face of the 'Pillar'. This row of stones was simply placed on top of the debris and was more in the nature of a demarcation line rather than a wall. In the south corner of the temple court a big, flat-topped, piece of white sandstone was put to use as a base for a round incense-altar, a type found also at Serabit el-Khadem. Both stones together were meant to be one *mazzebah*. Three more incense-altars of the Serabit el-Khadem type were found incorporated into the row of *mazzeboth* by being placed on rough pedestals of one or more stones piled on top of each other. Between these 'combined' *mazzeboth* stood simple standing stones in the form of long and narrow slabs of white sandstone. One of these standing stones had a pointed base set into a conical pit, cut some 25 cm. in to the bed-rock.

In the row of *mazzeboth* stood, obviously in secondary use, a square pillar with defaced representations of Hathor on two sides. This bears a strong similarity to the Hathor pillars at Serabit el-Khadem. Next to it was found a sandstone basin with a most unusual, heavy, granite boulder lying on top of it. The boulder, about 50 cm. high, showed no tool marks except for a flattened base and a partly flattened side, where apparently it had once been fitted against a wall. It was found lying on the basin and must have been intentionally so placed as it could not have fallen there from anywhere else. This unusually shaped boulder, the only granite object of its kind found in the excavation, must have had some special function in the original temple, which has yet to be established.

Plates 72, 78

Plate 69

Behind this rather impressive row of *mazzeboth* and all along the inside of wall 3 a considerable quantity of red and yellow cloth was found. The cloth was of a heavy kind, lying in a thick mass and in many folds, often with beads woven into it. A similar mass of folded cloth was found along the inside of wall 1, also outside the court in Locus 101, along and close to wall 1. The detailed study of these textiles, not yet concluded, shows that they consist of well-woven wool and flax of varying tints of yellow and red. The appearance of such large quantities of cloth, stratigraphically belonging to the last phase of the temple, and their location, *i.e.* all along walls 1 and 3, was at first hard to understand. It was obvious that they must have been part of the temple-furniture, some kind of hangings that had fallen down and been left lying where they were found. Yet we could see no structure on to which these hangings could have been attached. The problem solved itself when, during the clearing of the floor in Loci 107–109, two stone-lined pole-holes were found, penetrating into the white floor, but obviously not contemporary with it. These were the holes made to secure the poles of a large tent which, during the final phase of the temple, had covered the temple court. The temple had been turned into a tent-covered shrine, the first of its kind ever discovered. There are convincing reasons to relate this tent-sanctuary to the Midianites who seem to have returned to Timna for a short time after the Egyptian copper mining expeditions no longer reached the area, and worked and worshipped in their own way.

Plate 70

Fig. 44

The Midianites when refurnishing the Egyptian Hathor temple as a Semitic desert shrine cleared votive gifts, sculptures, inscriptions and Hathor sculptures out of the temple court. Most things were simply thrown out and piled behind the temple wall, out of sight of the visitors to the renovated shrine. This explains why most of the temple

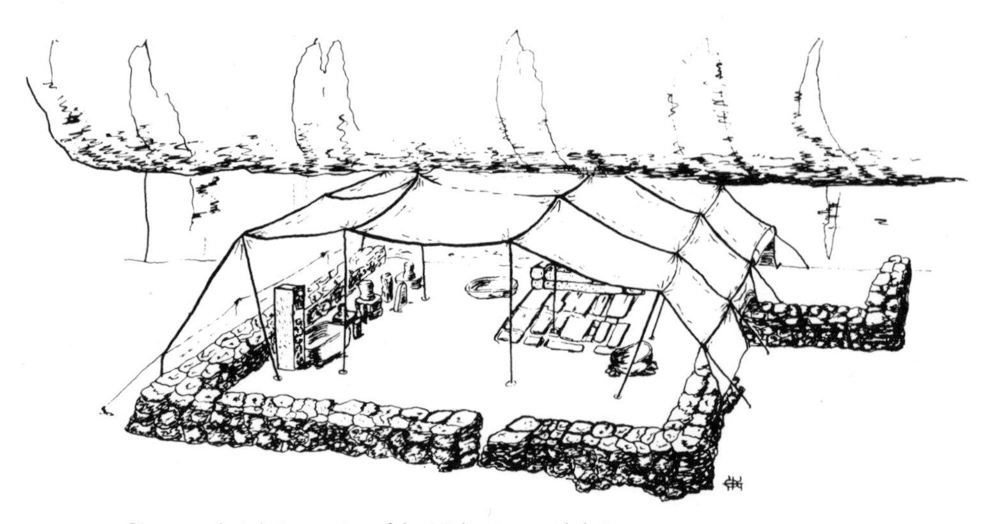

44 Site 200: Artist's impression of the Midianite tented shrine

gifts were found, broken and mixed up, in one thick layer in Locus 101. The level of this deposit, from about 25 cm. above the wall foundations to about 20 cm. below its top, *i.e.* very close to the present surface, corresponds well with the latest, Semitic phase of the temple. A number of architectural elements, the square pillars, altars, etc., were re-used in the renovation of the temple as building stones or as part of the *mazzeboth*, but care was taken to deface the Hathor representations and to erase any visible hieroglyphic inscription. The central niche in the naos was left empty, but the naos itself was re-used, apparently still the temple's most important part. Here, a Midianite copper snake with a gilded head was found *in situ*, the only votive object actually found inside the naos.

Plates XIX, XX

THE VOTIVE GIFTS

Almost ten thousand single small objects were registered from stratified contexts, painstakingly cleared from within the main strata II–IV of the temple. Together with the finds from strata I and V, the number of small finds at Site 200 is almost eleven thousand. The repeated and thorough destruction, repairs and rebuilding of the temple structure, ancient cleaning-out operations of whatever was lying on the floors and, foremost, the Nabataean metallurgists' intrustion into the temple ruins which had caused the utter destruction of any previously existing stratigraphy in the north-eastern half of the temple, made it best not to rely on the finds from strata II–IV for any chronological sequence. Therefore these finds are regarded as one group covering the whole period of the existence of the temple in all its phases, *i.e.* from the time of Sethos I to its terminating Midianite phase in the second half of the twelfth century BC.

Plate 71

THE POTTERY

On all but the lowest floors of strata II–IV were found the same three kinds of pottery which had occurred previously on the surface of the Timna sites and especially in the excavation of Site 2.

About 10% of the pottery found in the temple was of the primitive, hand-made Negev ware, containing as temper fragments of slag or of 'normal' pottery, and it seems to have been locally made. The makers of this pottery, who may have been Amalekite mining

Fig. 45

45 'Normal', wheel-made Late Bronze Age-Early Iron Age I ware (1–6, 13–24) and Negev-type, hand-made pottery (7–12) found in the Timna temple (Site 200)

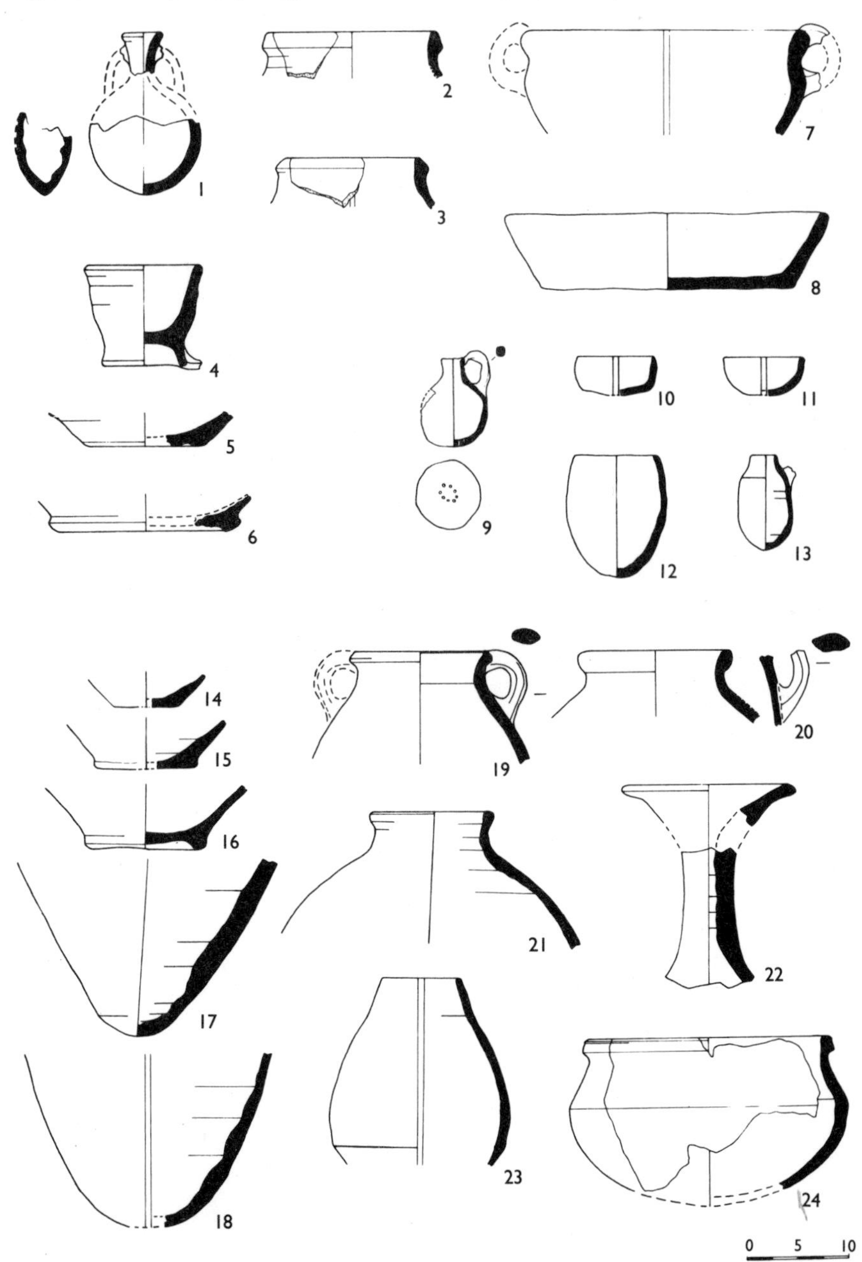

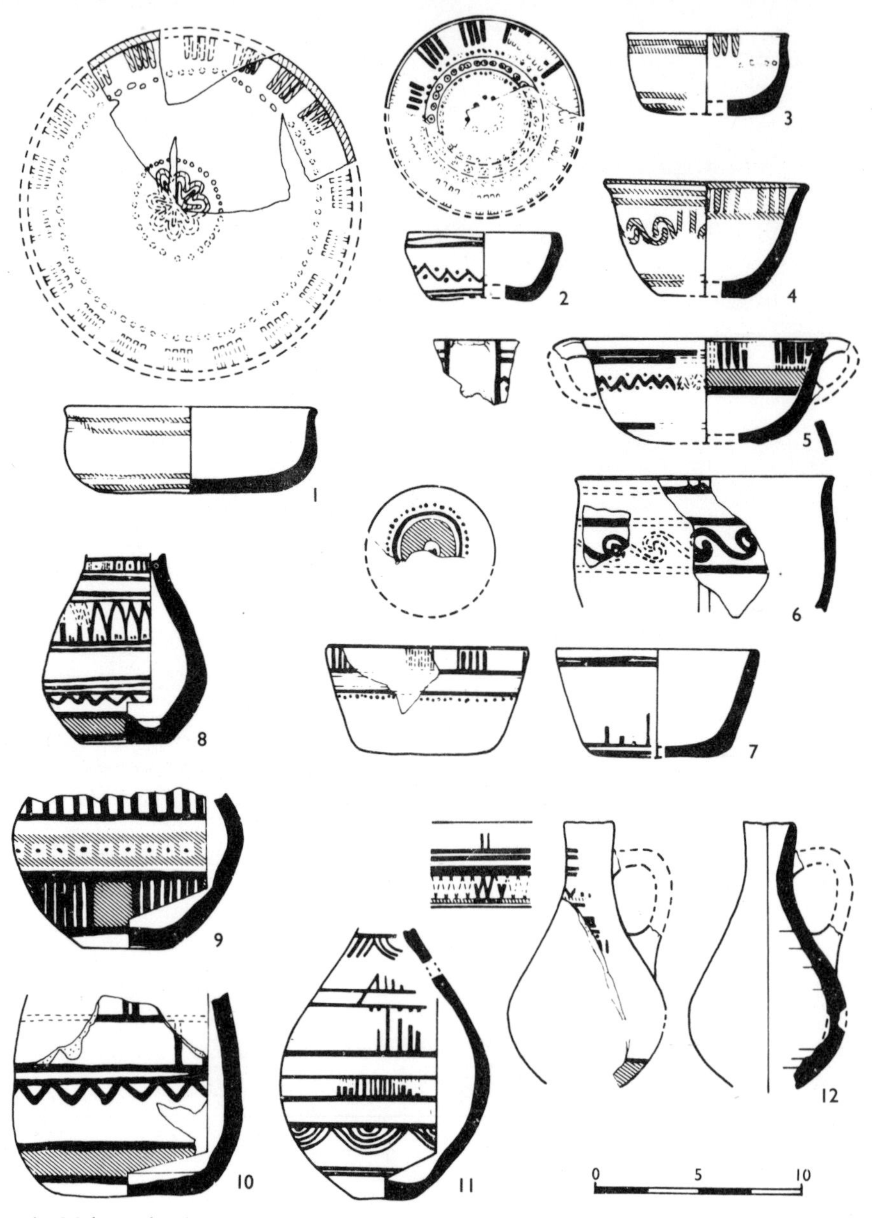

46 Midianite bowls, jugs and juglets from the temple site

workers recruited in the Amalekite settlements of the Central Negev Mountains, also imitated some vessels of the 'normal' wheel-made pottery used in Timna, but continued to use the same clay and slag temper whilst making their copies by hand. Also found were fragments of a primitive votive altar made of the same ware, with a snake crawling along its rim. Another new type of Negev ware was a small juglet with holes drilled through its bottom, which served perhaps as a strainer.

47 Midianite pottery decorated with ostrich drawings and strange human figure; 5 is a sophisticated incense vessel, all from the temple site

About 25% of the temple pottery was Midianite ware, of the same kind found on all the floors of Site 2, but this pottery did not appear in the temple in its initial phase. The Midianite pottery brought to the temple did not include any large, plain cooking pots as at Site 2, but consisted mostly of beautifully decorated small bowls, jugs and juglets. The decorations included large birds, probably ostriches, in sophisticated combinations with geometrical designs and also a human figure, with widespread fingers and strange head-gear,

Fig. 47

Plate XXIV; *Fig. 47, 4*
Plate 101

X Faience votive mask of the Egyptian goddess Hathor.

XI View of the Hathor Temple beneath the rock overhang under 'King Solomon's Pillars' (*see* Plates 65–73).

XII A selection of the thousands of carnelian, glass and faience beads found in the temple area.

XIII Egyptian amulets and beads of crystal, faience and shell from the temple.

XIV, XV Fragments of multi-coloured Egyptian glass of the XIX-XX dynasties.

XVI Two long glass beads, decorated with coloured glass thread ribbing, from the temple.

XVII A small primitive copper figurine.

XVIII A phallic idol, cast in Timna and of non-Egyptian origin (*cf.* Plates 102–4).

XIX, XX Midianite copper serpent with a gilded head. It was found in the central shrine of the Egyptian temple in its Midianite phase.

X

XI

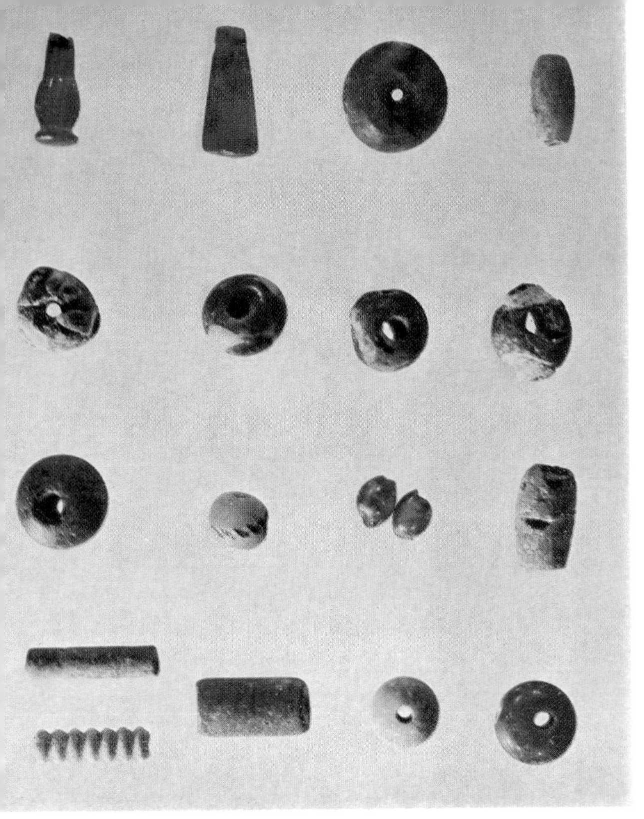
XII

XIII

XIV

XV

XVI

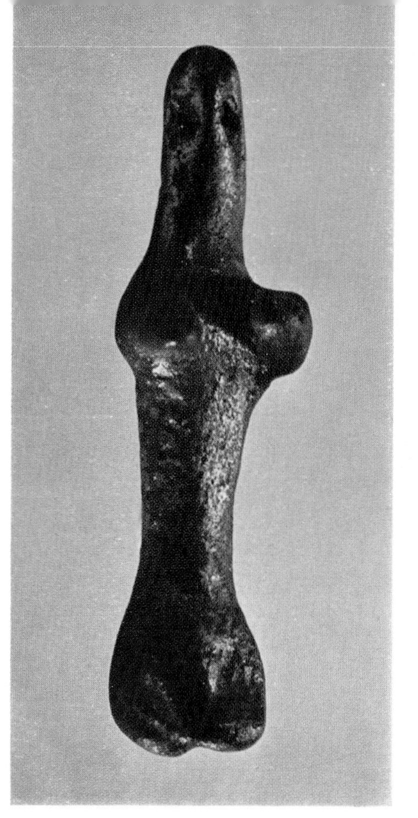

XVII

XVIII

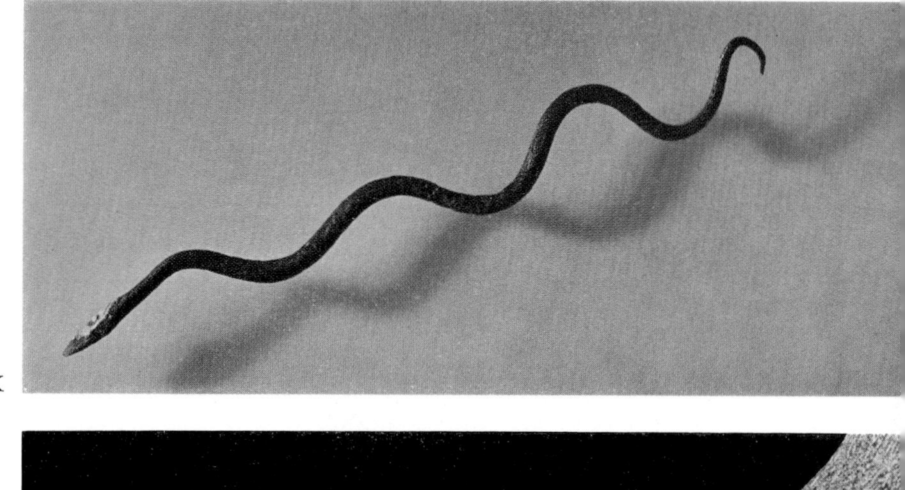

XIX

XX

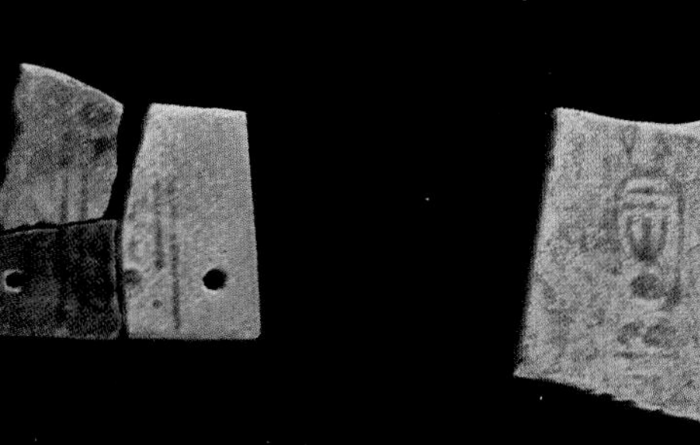

XXI

XXII

XXIII

XXIV

XXV

XXI Fragments of Egyptian faience *menats* with inscriptions in cartouches.

XXII Small Midianite bichrome bowl.

XXIII Fragments of Midianite bichrome-wares. Note the unusual use of red, black and white on the fragment, bottom right.

XXIV Part of a Midianite bichrome juglet with ostrich figure decoration.

XXV Blue faience amulet representing an Egyptian pharaoh wearing the Double Crown.

similar to some of the figures on Engraving 2 at Site 25. Also found was a unique vessel of superb workmanship, seemingly used for incense. It had the shape of a flat-bottomed cup with slightly indented walls, decorated with a series of birds, and a handle attached to the bottom consisting mainly of two thin, parallel, decorated fins running along the cup to its upper rim. The extraordinary variety and workmanship of the Midianite vessels found in the temple, as compared with the Midianite pottery from the other sites of Timna, is a clear indication that these sophisticated vessels were brought as votive gifts for the Hathor Temple. In the light of the finds of Midianite pottery with identical decorations on identical ware in north-western Arabia, it seems certain that the Midianite pottery was brought to the Timna temple all the way from there, perhaps from the large Midianite town at Qurayyah, about 160 km. south of Aqaba.

Fig. 47, 5

The remaining 65% of the pottery found in the temple was 'normal', wheel-made Late Bronze Age–Early Iron Age ware. In addition to the pottery types found also at Site 2, several new vessels made their appearance in the temple, together with some small flasks and two incense-stands.

Fig. 45

The realization that the Timna copper works, with a Hathor Temple in its centre, was in fact a Pharaonic industrial undertaking dated to the New Kingdom brought the problem of the provenance of the pottery found in Timna to the foreground. Whilst sufficient comparative evidence existed from the Central Negev Mountains, the Arabah and north-western Arabia to relate the Negev ware and the Midianite pottery to the areas inhabited by the Amalekites and Midianites, the third and largest group of pottery, the 'normal', wheel-made Late Bronze Age–Early Iron Age pottery cannot yet be traced to its origin or its originators. Although most of its forms and shapes can be compared to Palestinian pottery of the period it seems strange to find so much Palestinian pottery in an Egyptian industrial undertaking in the southernmost Arabah, operated in collaboration with tribes from the Negev Mountains and, more so, from Midian. There also exists written evidence (Harris Papyrus I) for direct supply connections with Egypt. Therefore the possibility that the 'normal' wheel-made pottery of Timna was imported from somewhere in Egypt was investigated.

Petrographic analysis of the Timna pottery proved highly interesting, though not entirely conclusive in regard to the problem. It showed that some of the Negev pottery contained as temper tiny fragments of the fayalite slag from the smelters of Timna as well as crushed sherds of the rather distinctive 'normal' Timna ware. Its clay is red, very fine and isotropic. There can be no doubt that at least some of the Negev ware was made in the Timna Valley. The Midianite pottery differs in its clay and temper from any known pottery of Syria–Palestine. The colour of the very fine clay is a pale beige, it is homogenous, anisotropic and contains much mica, quartz and iron oxides. Its temper consists of tiny red fragments, probably of finely crushed burnt bricks. The 'normal' wheel-made pottery is not as homogenous as the other two groups. One such shered contained as temper fragments of magmatic and metamorphic rocks, whilst the others all contained quartz sands and fragments of various pottery. The clay of all the 'normal' sherds contains distinctive particles of rhomboidal shape, perhaps dolomite or selenite. The mineral content of this pottery does not allow the accurate identification of its source but it is certain that it originates from somewhere in the semi-arid or arid areas of the pre-Cambrian massif, spreading from the eastern Delta of Egypt, through Sinai and the Arabah and well into Arabia. In other words, the 'normal' wheel-made pottery found in all Late Bronze Age–Iron Age I sites of Timna could have been brought to the Timna copper works from any supply centre in the area adjoining the large desert east of the Delta, but does not seem to come from Egypt itself. Unfortunately, there is no comparative material available from this area and the quest for the origin of the 'normal' wheel-made Timna pottery must remain, for the present, unanswered.

THE EGYPTIAN OFFERINGS

A considerable number of Egyptian-made votive gifts were actually brought to the Timna Temple by the Egyptian mining expeditions. Of the greatest importance among these offerings are the objects bearing hieroglyphic inscriptions and cartouches containing the names of Egyptian pharaohs (studied by Dr R. Giveon). There were a number of faience *menats*, bearing the names of Ramesses II and Ramesses IV, and a small, blue faience seal together with a sistrum handle carried the name of Sethos I. The cartouche of Ramesses II occurs on fragments of a faience bowl, a fine green glass fragment, and on a number of bracelets. Another bowl bears the cartouche of Merneptah. The name of Ramesses III appears on a ring stand, and Ramesses IV on a *menat* fragment, on a beautiful lotus-decorated bowl and on several bracelets. The names of both Sethos II and Ramesses V were also found on bracelets.

Plate XXI
Plates 82, 83
Plate 85
Plate 84

The Egyptian kings thus represented by inscriptions in the Timna Temple are: Sethos I (1318–1304 BC), Ramesses II (1304–1237 BC), Merneptah (1236–122 BC), Sethos II (1216–1210 BC) and Queen

Fig. 48

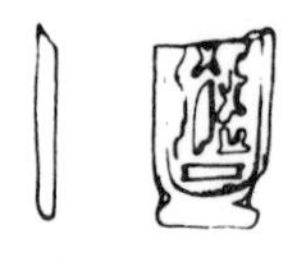

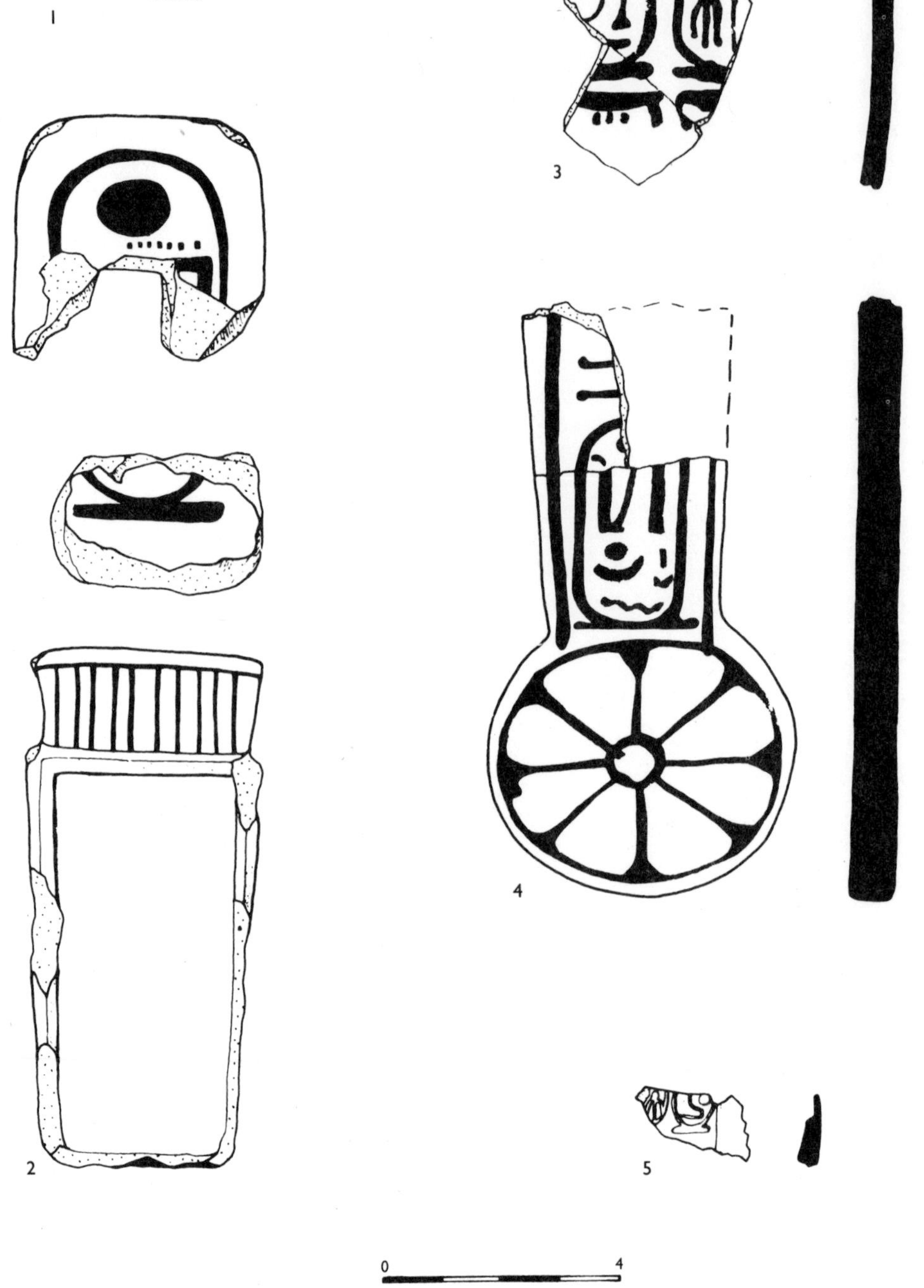

48 Cartouches found in the temple containing the names of Egyptian pharaohs: 1–2 Sethos I; 3–5 Ramesses II

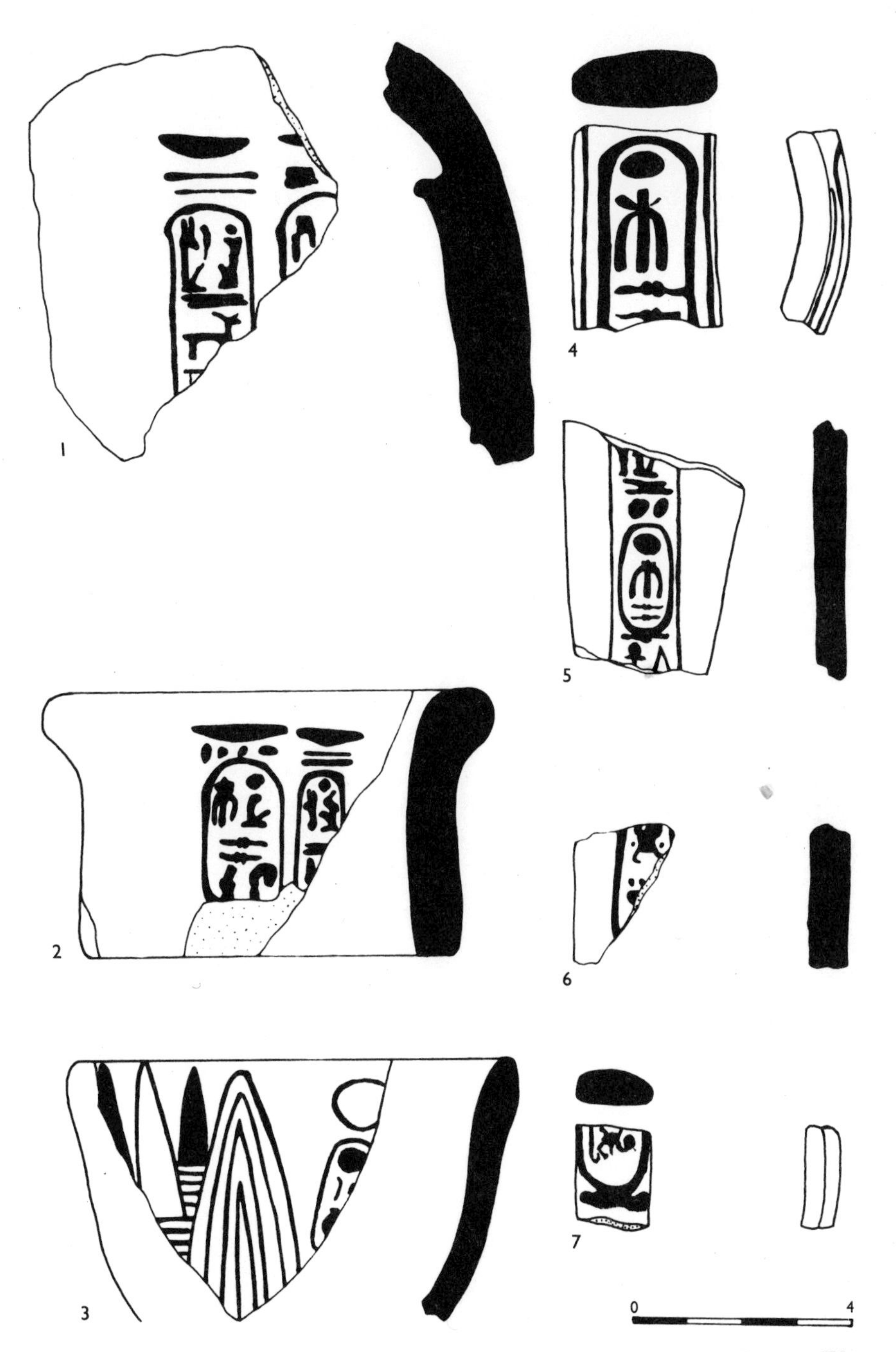

49 Egyptian cartouches from the temple site: 1 Merneptah; 2 Ramesses III; 3–5 Ramesses IV; 6 Sethos II; 7 Ramesses V

Fig. 49

Twosret (1209–1200? BC). These all belong to the XIXth Dynasty. From the XXth Dynasty there are: Ramesses III (1198–1166 BC), Ramesses IV (1166–1160 BC) and Ramesses V (1160–1156 BC).

Of particular interest is an inscription found on a bracelet fragment reading, 'Hathor, Lady of the Turquoise', together with a similar one on a sistrum handle. 'King of Upper and Lower Egypt' occurs on several bracelets. The lower, inscribed half of an ushabti figure of blue faience was also found but, unfortunately, the remaining text does not include the owner's name.

Among the uninscribed faience offerings were fragments of vases and bowls, some beautifully decorated with spirals, wave patterns and lotus flowers. A very fine bowl shows a fish amid lotus flowers and plants and there were fragments of other fish bowls. Several wands were found, bearing the *wedjet* sacred eye on both sides. Various heads of Hathor, made of glazed pottery, were in fact sistrum handles, with a face of the goddess on each side. A number of other sistrum fragments were also found but none would join any of the pieces found in the temple on most of which Hathor has pronounced cow-ears.

Among the votive offerings especially connected with the worship of Hathor at Timna are representations of the cat family (*felidae*) including one leopard (*Panthera pardus*). Cats appear as drawings on flat tablets, just as at Serabit el-Khadem, but mainly as figures in the round made of blue faience. A fine amulet of Bastet, made of a thin piece of bone, and a faience amulet of the same goddess should also be mentioned here as being related to Hathor worship. Numerous representations of the god Bes in the form of small phallic, magic amulets were also found. Another faience amulet possibly represents a king wearing the Double Crown of Upper and Lower Egypt. A number of scarabs and seals were found, bearing animal representations, geometrical designs and hieroglyphs. There were also some alabaster objects, including a complete small vase, and the foot of a small statue. Only a few metal objects could be identified as of Egyptian origin; a finely incised gold head-band, a silver button, some tiny gold beads and various metal parts of sistra. A wooden comb, still in excellent condition, was of considerable interest. According to A. Fahn, it was made of *Buxus sempervirens*, a wood still used in many countries for the manufacture of combs, and which grows in Europe, North Africa and Asia but not in Egypt or Palestine.

There were numerous fragments of carved stone objects and statues, some quite small, others parts of large architectural elements. Stone was also used as raw material for bowls and decorated jar-lids

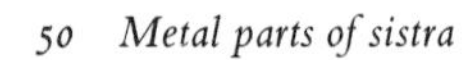

50 *Metal parts of sistra*

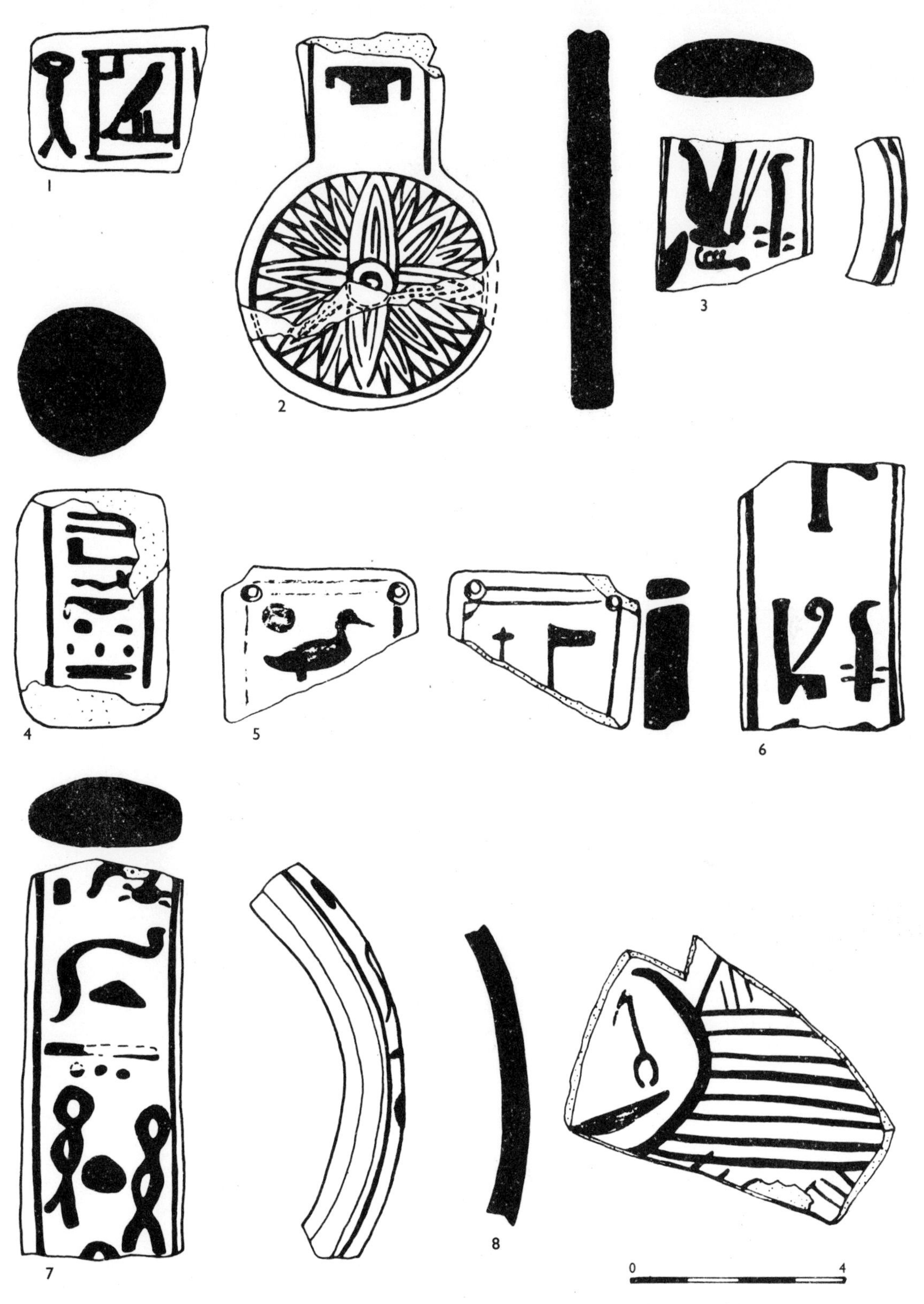

51 Inscribed Egyptian faience votive gifts from the temple (read by R. Giveon): 1 'Hathor Lady of the Turquoise'; 2 Sign for 'gold', originally written after a king's name; 3 'King of Upper and Lower Egypt'; 4 '[King N] beloved [of Hathor] Lady of the Turquoise'; 5 Left: 'Son of Re'; Right: 'the Beneficent God'; 6 'King of Upper and Lower Egypt'; 7 '[Given life] like Re for ever eternally'; 8 Bowl fragment with the blessing: 'Lord of Life, Stability, Well-being'

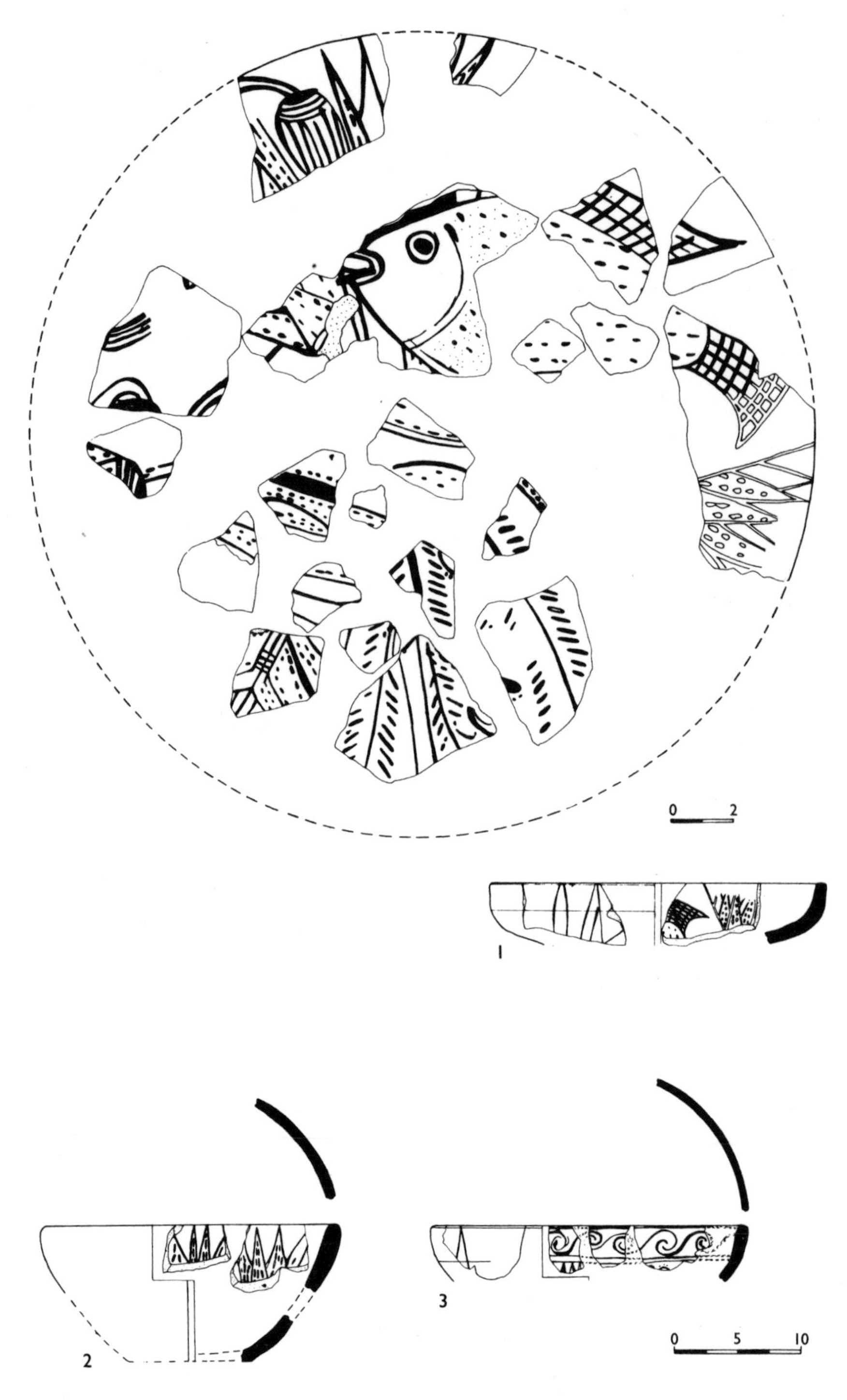

52 Faience bowls with fish and lotus flower decorations from the temple

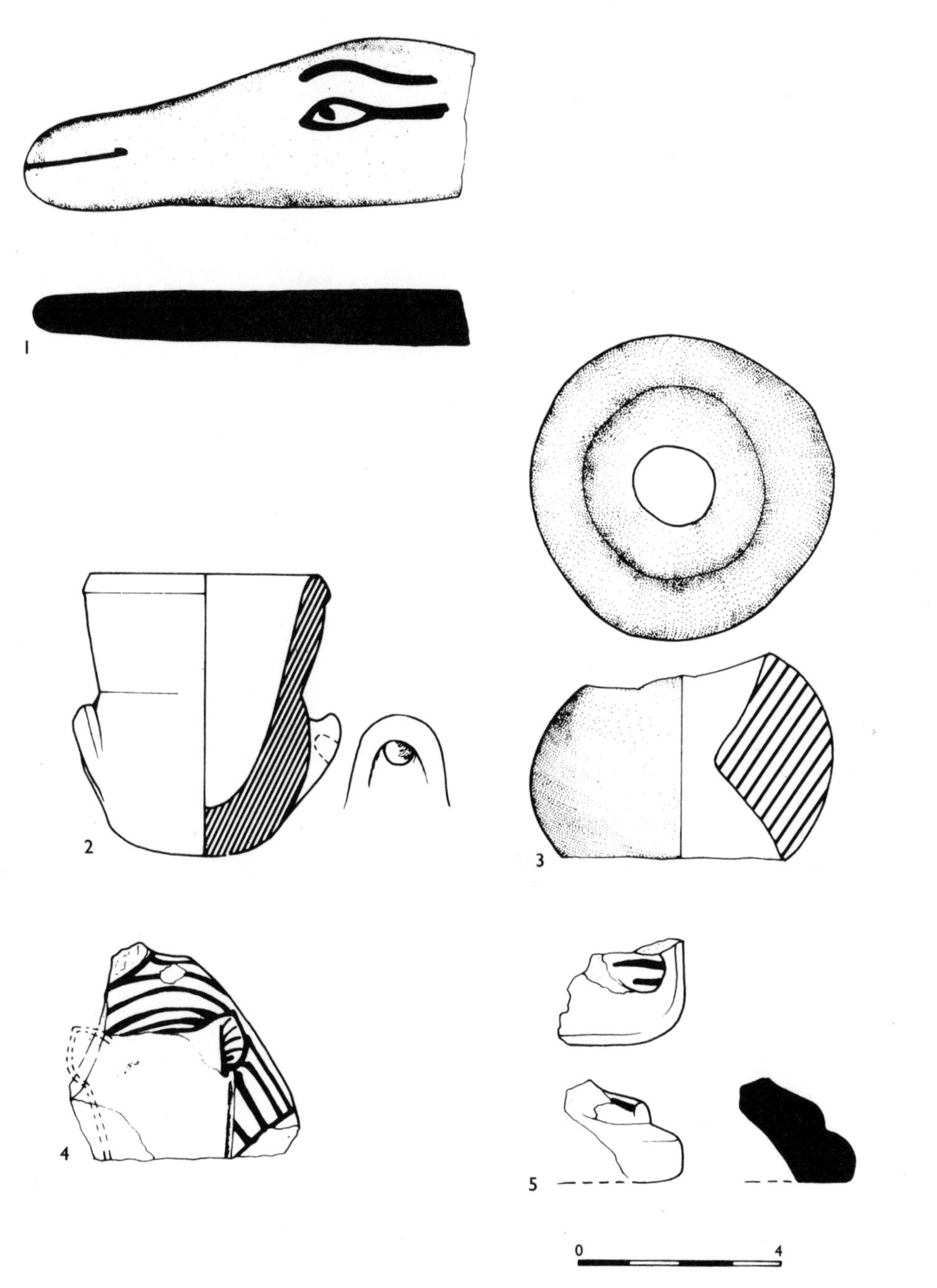

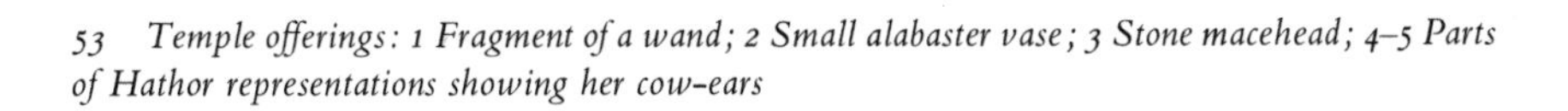
53 Temple offerings: 1 Fragment of a wand; 2 Small alabaster vase; 3 Stone macehead; 4–5 Parts of Hathor representations showing her cow-ears

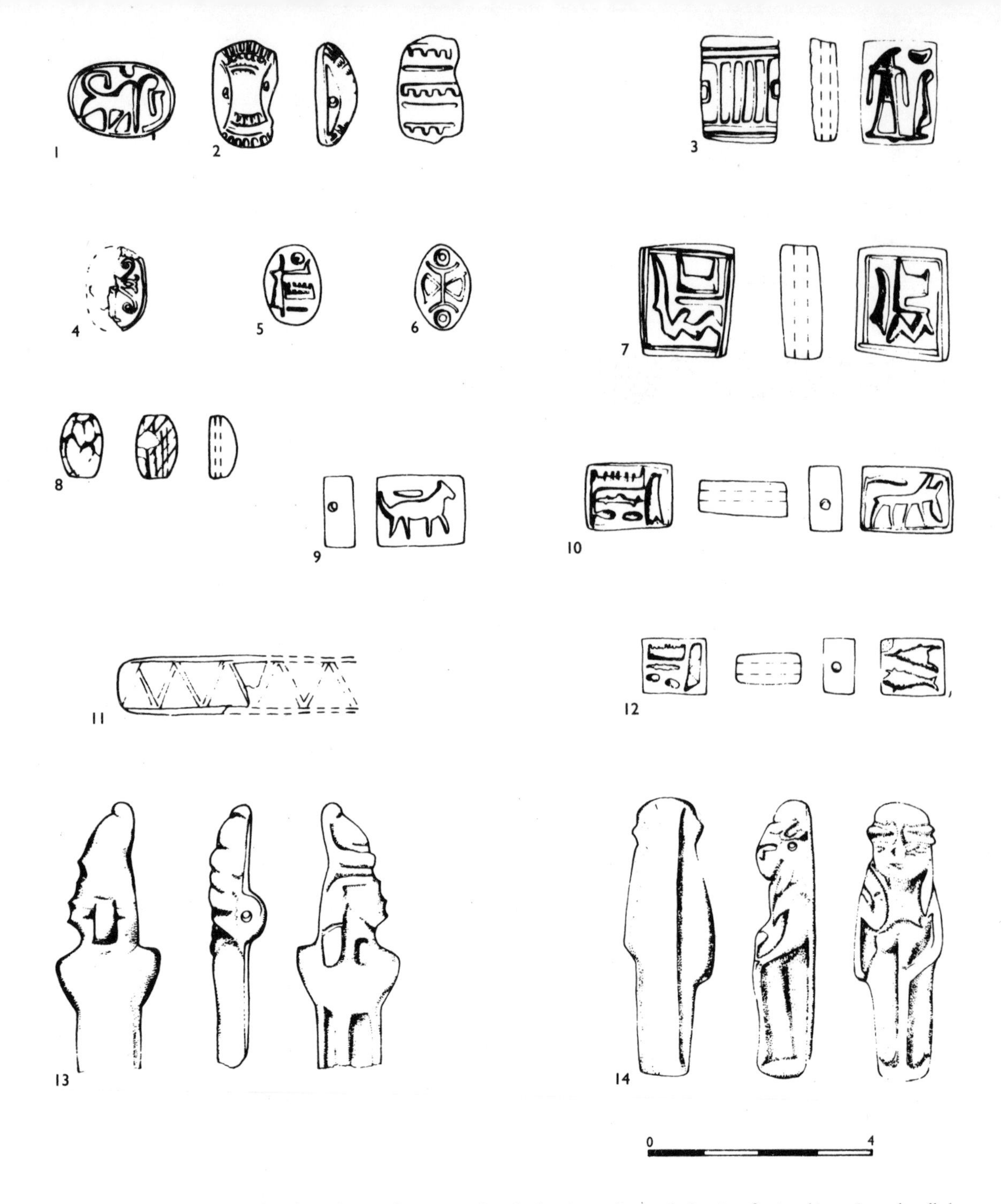

54 Scarabs, seals and amulets from the temple area: 1 Scarab showing a lion and the sign for 'truth'; 2 Loop-handled faience seal with geometrical pattern; 3 Hemi-cylindrical faience seal with falcon-headed solar deity and the sign for 'truth'; 4 Faience scarab; 5 Scarab; 6 Faience scaraboid with geometrical pattern; 7 Clay seal, both sides reading 'Amon-Ra'; 8 Small faience seal; 9 Primitive seal with unidentified animal; 10 Double-sided seal; 11 Gold band; 12 Double-sided faience seal: two scorpions and 'Amon-Ra'; 13 An unidentified king wearing the Double Crown of Egypt; 14 Faience figurine of the goddess Bastet

as well as for gaming boards. A 30 cm. high rectangular podium of local white sandstone had the lower half of a cat or sphinx sitting on it but very many of the stone objects were defaced beyond recognition.

Plate 75

Numerous fragments of Egyptian glass were found in all parts of the temple and made the subject of a special study by Miss G. Lehrer. There are over 150 fragments from 50–60 core made vessels and also fragments from a flat, thick inlay piece. This group is of particular importance because of its closely dated context with XIXth and XXth Dynasty inscriptions. The following types could be identified: krateriskoi, lentoid and globular flasks, vessels with elongated body and rounded base (amphoriskoi?) and bowls. With the exception of four fragments of green glass, all the vessels have a background either of light blue, mainly an opaque sky-blue, or of dark blue in shades of cobalt or dark turquoise. The thread decorations of garlands or festoons, zig-zags and feather pattern are white, yellow and dark blue on the light blue backgrounds and white, yellow and light blue on the dark blue backgrounds, applied either in one colour, or a combination of two or three colours together. A common feature of these vessels is the twisted black and white strip applied on almost every rim and in several places also on the shoulder and lower parts. Several fragments of opaque light blue bowls have yellow thread applied on the rim. Fragments of green glass include a rounded base, part of a vessel with the double cartouche of Ramesses II, part of a bracelet with traces of an impressed pattern and a fragment possibly from the rim of a pomegranate bottle. This whole group belongs, from the point of view of typology and decoration, to the end of the XVIIIth and to the XIXth Dynasties.

Plate XIX–XVI

More than 5000 beads, parts of necklaces, collars, pectorals and bracelets, were found and studied by Mrs T. Kertesz. Many of the beads were made locally mainly as disc beads of mica schist, small pebbles and Red Sea shells, perforated and strung as necklaces. These beads were probably brought to the temple by local workers. The main body of beads brought from Egypt consisted of more than 2500 faience discs in blue, green, brown and white, some gadrooned and shaped to resemble the daisy flower. There were about one thousand short beads of faience, around one hundred made of carnelian or limestone, and some thirty of coloured glass. Many beads of standard circular shape were made of glazed faience or frit, limestone, carnelian or onyx. Most beautiful are the decorated glass beads made by the 'eye spot' technique: drops of yellow, brown and blue glass were inserted into the white or black matrix of the bead.

Plates XII, XIII, 98

Sometimes a differently coloured ring was laid around the eye spot or stratified eyes were superimposed on top of each other. Other glass beads were made by the 'composite coil' technique: two differently coloured glass bands, white and black or white and red, were wound around the yellow or brown core of the bead. There were also many beads of long tubular or barrel shapes, made of faience, carnelian and glass. The long carnelian barrel beads, called in Egypt *sweret* beads, had special magic properties when worn on a thick, twisted cord close to the throat. Two beads of glass, 50 mm. long and 20 mm. wide, with multi-coloured glass thread wound around their bodies are the largest found in Timna. There were also faience amulets shaped like open lotus and papyrus flowers, a carnelian pendant shaped as a lotus bud and numerous other pendants and amulets of haematite, mica, quartz, bone and gold leaf.

Plate XVI

The whole group of Egyptian offerings found in the Hathor Temple of Timna shows strong similarities, even down to minor details, to the offerings found by Sir Flinders Petrie in the Hathor Temple of Serabit el-Khadem in Sinai. Added to which, inscriptions reading 'Hathor, Lady of the Turquoise' or '. . . beloved Lady of the Turquoise' on faience offerings found in Timna, the valley of the copper mines, being identical with inscriptions found at Serabit el-Khadem in the temple of the Egyptian turquoise miners, seem to indicate a central organization for the preparation and supply of the Egyptian mining expeditions into the desert.

MIDIANITE AND OTHER LOCALLY MADE OFFERINGS

Besides the beautifully decorated pottery of Midianite origin, very numerous votive gifts made of metal were not imports from Egypt. Whilst there is some evidence, such as the style of decoration, for a Midianite origin for some of the metal offerings we, nevertheless, have to rely in most instances on the non-Egyptian character of the finds as a pointer to their most probable, Midianite source. In many cases the metal objects are obviously locally made and in the light of the overall picture of the excavation one may relate them to the Midianites in certain instances and in others to the Amalekites, the contemporary inhabitants of the Central Negev Mountains and the Arabah. The fact that nothing like the metal offerings attributed to the Midianites was ever found in the Hathor Temple at Serabit el-Khadem, where similarly no Midianite pottery exists, seems a strong argument for the proposed ethnic identification. Yet, it must be said that a certain degree of reliance is placed on Biblical traditions relating to the Kenites-Midianites as the ancient metal-workers of the southernmost Arabah and in the area of the Red Sea. The majority of the Midianite temple offerings were locally made or

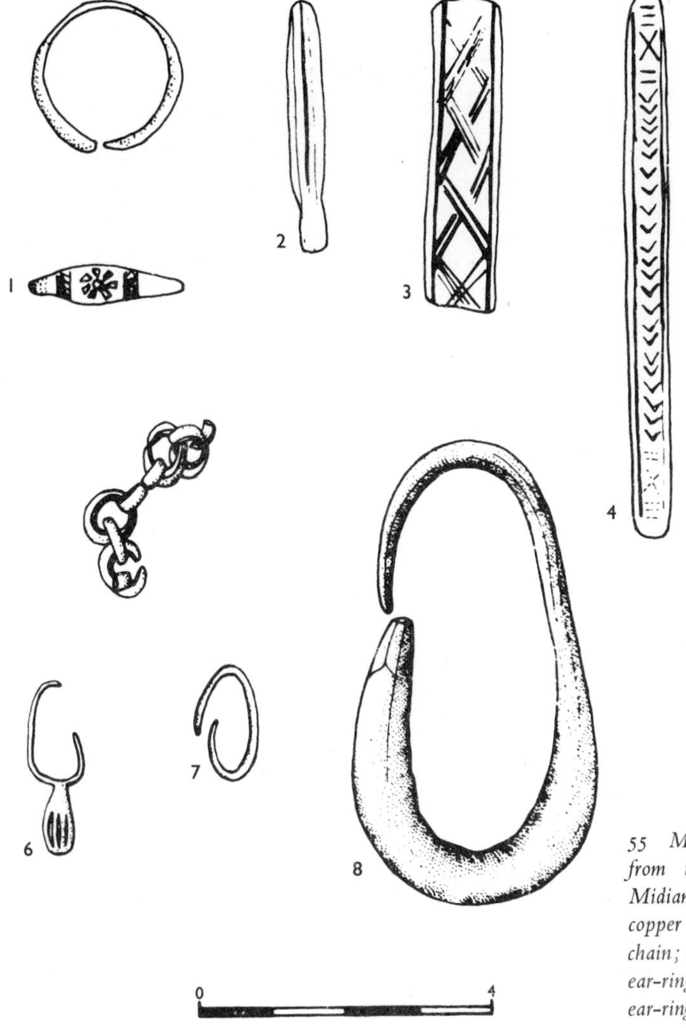

55 Metal votive gifts from the temple: 1–4 Midianite decorations on copper rings; 5 Copper chain; 6, 7 Gilded copper ear-rings; 8 Large copper ear-ring (42 gr.)

each side. It could well be the beam of a balance and the chain fragments found in the temple may belong to it. There was only one spear-head and one arrowhead, the latter found on the surface of the site. Several copper ferrules with rivet holes, and some with the actual rivets still in place appear to have been attached to a handle, but no traces of the tools themselves were found.

Fig. 58

Fig. 59

Of considerable interest is the appearance of iron in a thirteenth-twelfth century BC context. Previously, in 1964, two iron bracelets had been unearthed in Site 2, Area F, but now there are in addition several well-made iron rings, one with clear remains of gilding. Iron was obviously still a rare material used only for jewellery.

Plate XVII

The list of metal offerings ends with two unique figurines with human features. One is a 4 cm. high, primitively cast, copper figurine, imported copper-based metal objects, like jewellery and cosmetic equipment, figurines, and very few weapons and tools. A large number of fragments of metal objects have not yet been identified or explained. Much analytical work is still to be done on the metal finds from Timna.

The most significant and remarkable find, already mentioned above, was a lifelike copper serpent with a gilded head and a finely worked, smooth body, unearthed inside the naos and belonging to the final Iron Age I phase of its history. It is 12 cm. long and represents a colubrid snake of the racer type. Its gilded head is finely shaped and shows two large eyes.

Plate XIX

Plate XX

A cast and finely polished copper figurine of a sheep (*Ovis aries*), about 4 cm. long and 2·5 cm. high, with heavy, twisted horns, was found. It had a hole through its neck and may have been an amulet. This little figurine is an exceptionally fine piece of metalwork.

Plate 97

More than one hundred copper rings were found, mainly from the last phase of the temple, in the copper hoard and in the 'refuse' pile unearthed next to wall 1 outside the temple court. Most of these are made of thin, narrow copper strips, bent into the shape of a ring and sometimes decorated with geometrical incisions resembling the decorations on the Midianite pottery. Other rings were made of copper wire, showing a round and/or rectangular profile.

Fig. 55

Another typical votive gift is the ear-ring, present in many variations and sizes. Some are simply oval-shaped wires, open on one side, others are cast into shape with a slightly thicker lower part. Some have a small pendant attached, simply decorated by incised lines. One example, 6·2 cm. long and weighing 42 g., was not, of course, intended for actual wear in a human ear, but seems more suitable for the large cow ears of Hathor.

There were also numerous examples of toilet and make-up equipment such as spatulae of various sizes, decorated sometimes with an incised 'tree-of-life' or the five fingers of a 'hand'. The 'hand' was a common decoration, probably with a magic intent, and it also occurred on small copper amulets made of thin sheet of copper. Toilet pins were made of round, square or rectangular copper wire, bent on one side into a loop-shaped handle. Very many twisted and bent wire-made objects were found, which may have been hooks or pins, perhaps used for strengthening or holding together the tent covers of the Midianite shrine. Many twisted wires were actually knots.

Fig. 57

One very interesting copper object consisted of a square beam, 18 cm. long, with both ends shaped into a loop with a small ring on

Fig. 56

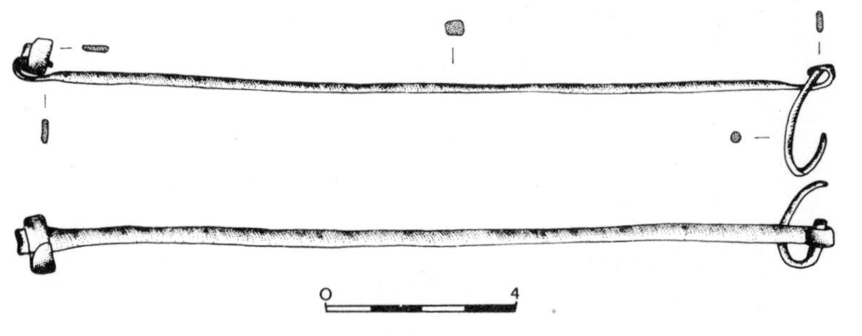

56 Balance beam with attached chain-link

57 Metal votive gifts: 1–5, 8–10, 12 Make-up implements and spatulae; 11 Spearhead (?); 6–7 Copper amulets with 'hand' decoration

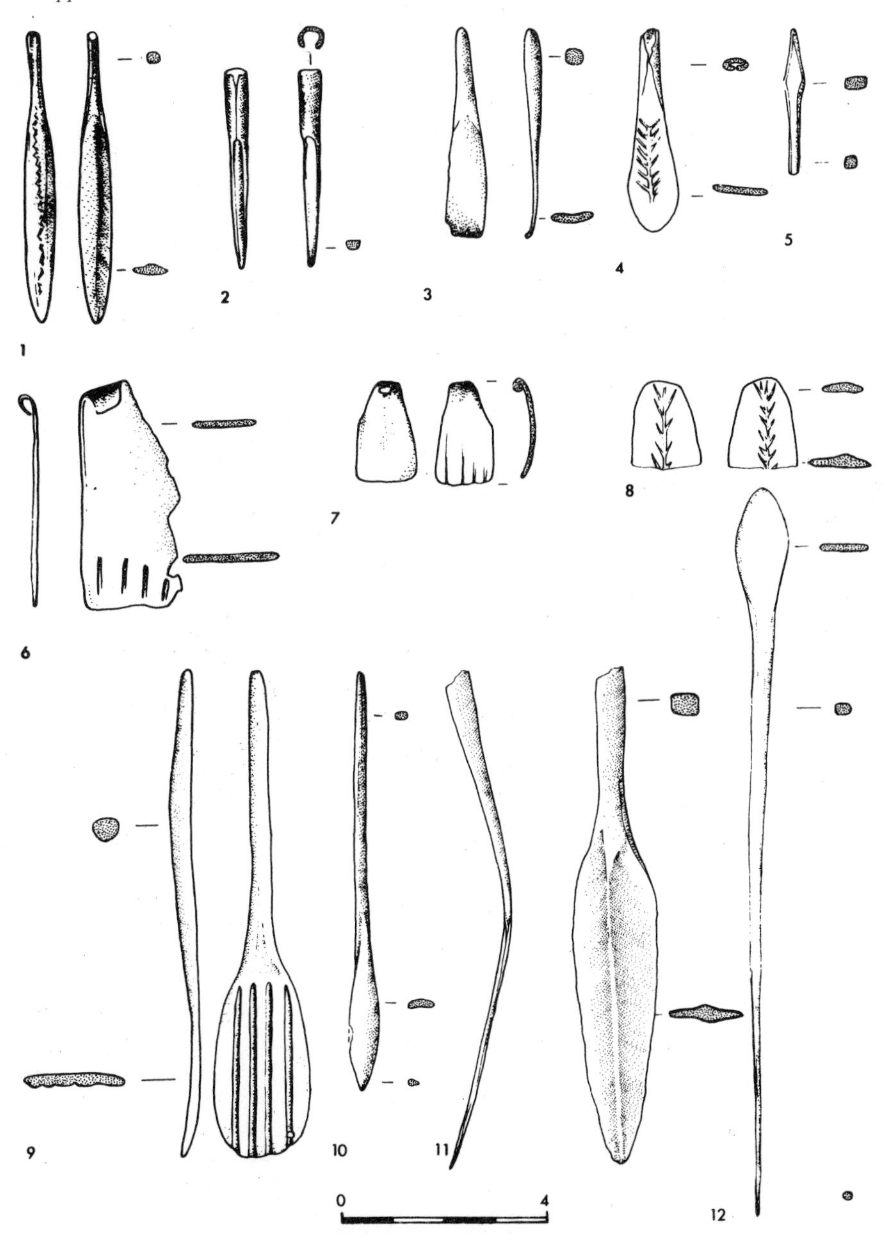

58 Copper ferrules showing riveting

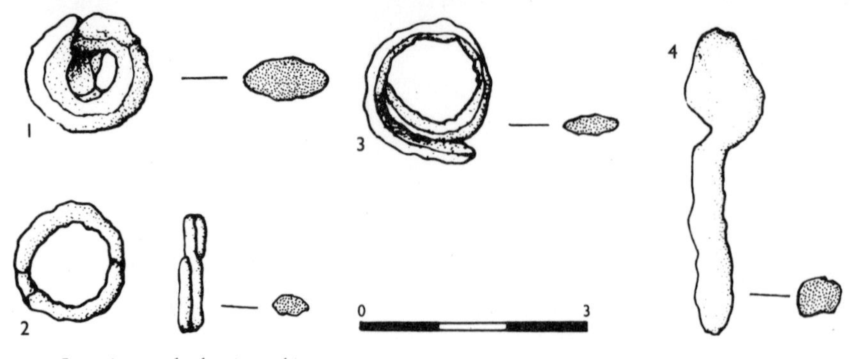

59 Iron rings and other iron objects

its short arms stretched forward and the upper part of its legs serving as the base of the figurine. There are no feet, but a phallus is indicated, and instead of a face two tiny notches indicate the eyes. The figurine appears to be sitting on a flat seat.

Plate XVIII

The second figurine represents a male with a rather emphasized phallus, a bearded, gross, face and primitively shaped arms and hands pressed against its sides. It has large ears and seems to wear some kind of head-dress. The features of this figurine are fairly well shaped but it was found 'as-cast', with the seams of the mould clearly visible. In fact, part of the mould was still stuck between its legs and it therefore seems obvious that the figurine was cast in Timna. No parallel to this figurine is known and it can only be suggested that it is perhaps a magic fertility symbol, or god, of non-Egyptian origin.

Plates 102–4

Besides the large number of votive artifacts brought to the temple as offerings to Hathor, numerous strange and curious objects were found in the temple which evidently had caught the imagination of the miners and were therefore brought as gifts to the mining goddess. Mention has already been made of the large number of perforated shells found used as beads, which were most probably brought to the temple by the Midianites, the inhabitants of the Red Sea coast. Many more beautiful shells and sea-stars, pieces of coral and strangely shaped pebbles and stones were found in the temple, some resembling mother-and-child figurines or large breasted women. There were also very many fossils, still to be found plentifully in the Middle White horizon of Timna, which looked like female figures and were also brought as offerings. A number of very rough copper casting waste pieces in the shape of animals were also offered as gifts to Hathor, though it is quite possible that some of these objects, such as a four-legged animal or a horse-and-rider, were actually the results of intentional primitive castings.

Plate 108

Plates 105–7

Plates 99, 100

A considerable quantity of bone fragments was found, mainly around the *mazzeboth*. Conspicuously there were only young goats, which could well be remains of votive sacrifices.

A Nabataean casting installation in the Temple Court

Many hundreds of years after the abandonment of the Hathor Temple, in the Roman period, when the temple site had turned into a low hillock of sand with several huge sandstone boulders which had fallen down from the mountain above lying on top of its north-western corner, a small group of apparently Nabataean metalworkers came to the site. The upper part of wall 1 and some of the eastern end of wall 2 was then still above ground and they swept away the debris from Loci 109 and 110 and settled into the cleared space. During this clearing operation the newcomers reached the previous white floor and also the remains of the naos. One of the results of this was an accumulation of debris in Locus 101 mixed with a large number of the small finds, described above, from the previous temple. Another result was a mixed layer of finds on the white floor, including Nabataean-Roman sherds together with cartouches of the Ramesside period and many other Egyptian and Midianite gifts.

Fig. 60

In Locus 109 a crucible melting furnace was constructed of white sandstone, taken from the debris of the temple. It was a square structure, 80 × 80 cm., containing a small hearth, 30 × 30 cm., open to the east. The furnace bottom and part of its structure were found *in situ* in the excavation. A pit, perhaps for charcoal, was found next to the furnace. A second fireplace, probably also a crucible melting furnace, was found next to wall 2 within a very thick layer of wood ash. In fact, a 20–30 cm. thick layer of wood ash was piled into the corner of walls 1 and 2 and covered most of Locus 109 up to wall 5. Wall 5 was constructed from previous temple building stones and the basin of soft white sandstone was re-used here. In the corner of walls 5 and 1 a very interesting store of goods was found, obviously put there by the Nabataean metalworkers: about 50 kg. of very rich copper ore nodules and numerous pieces of fossilized tree (iron oxide flux), mixed in complete disorder with many fragments of copper implements as well as broken and complete copper rings and copper wire, and copper pellets extracted from slag heaps somewhere in Timna. Ore and flux were not used at the temple site itself, as no evidence of smelting was found there, and it must be assumed that it was only collected by the Nabataean metallurgists and stored to be transported to a smelter elsewhere. The metallic copper, however, was melted in the crucible furnaces found in the temple court and cast into objects or ingots. Several fragments of slagged crucible fragments, as well as some casting slag, were found in Locus 109.

Plate 71

It is interesting to notice that there were no votive gifts amongst the copper objects collected by the Nabataean metalworkers for re-

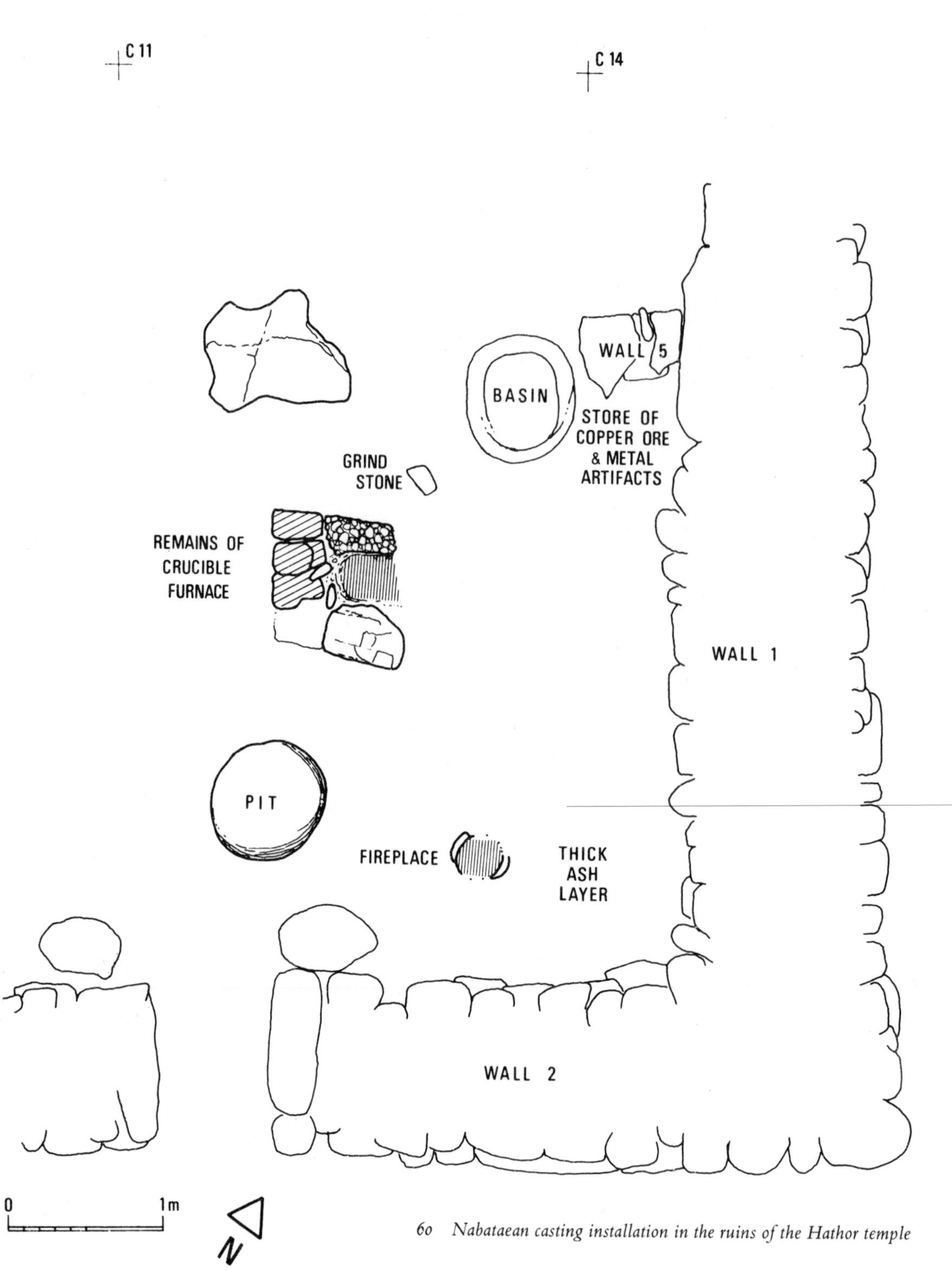

60 *Nabataean casting installation in the ruins of the Hathor temple*

smelting, only plain copper objects. The votive objects were discarded, either simply thrown over the wall or collected and hidden outside of wall 1, where the hoard was found at the beginning of the excavations. Whether the Nabataeans did so out of veneration for the ancient temple or for other reasons is a matter of conjecture, but the fact as such is worthy of note.

Plate 95

Greenish and light brown glass fragments from small bottles were found together with Nabataean-Roman pottery. They are free blown and, in one case, mould blown with a ribbed pattern. These glass bottles date to the first century AD. A fair quantity of Nabataean Roman pottery from the site included decorated oil lamps, juglets, jar fragments, a cup, piriform unguentaria and several shallow bowls. Some of these sherds were also found in Locus 111, above a layer of debris and wind-borne fill. No sherds of the period were found anywhere in the temple except in the area used by the Nabataean metalworkers, *i.e.* Loci 109, 110, 111, and also 101. The Nabataean pottery, found also in the waste from the melting furnaces, dates from the first century AD.

Fig. 62

The Nabataean casting installation at Site 200, was not an important industrial undertaking but rather a temporary establishment for the exploitation of the large Egyptian slag heaps in the Timna valley. The Roman copper industries of the Arabah are described pp. 208 ff.

61 *Nabataean-Roman pottery of the first century AD from the temple site*

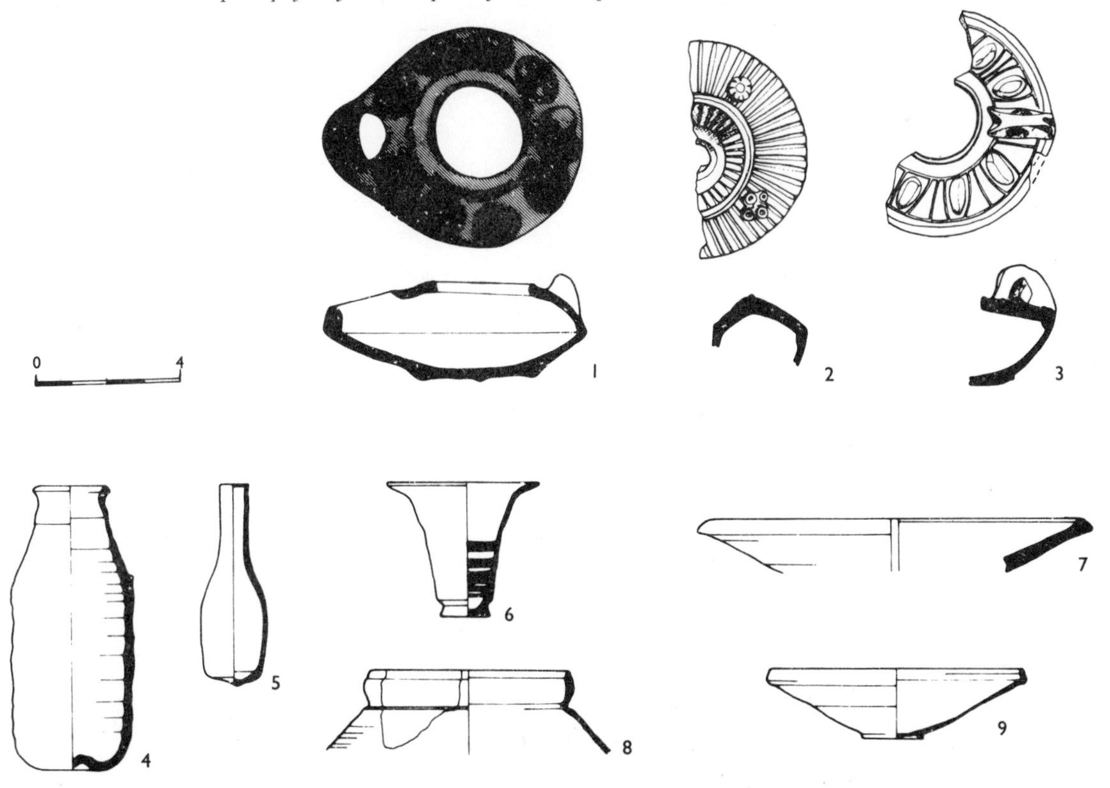

VI

The Hathor Temple and its Implications

The discovery at Timna of a temple dedicated to Hathor and dated by inscriptions to the XIXth and XXth Dynasties of Egypt, together with the fact that the pottery found in this temple is the same as the pottery found in the excavation of Site 2 and on the surface of all Late Bronze Age–Early Iron Age sites in the Arabah, finally ended the protracted discussions about the date and originators of the Timna copper industries. We know today that the Timna copper mines and all contemporary copper mines on the west side of the Arabah and in the Mountains of Elat belong to the period between the end of the fourteenth century BC and the middle of the twelfth century BC and were operated by Pharaonic expeditions of the XIXth to XXth Dynasties.

There is no evidence whatsoever of any copper mining or smelting activities in the western Arabah later than the twelfth century BC until the renewal of the industry in the Roman period. There is no factual and, as a matter of fact, no ancient written literary evidence of the existence of 'King Solomon's Mines'. More so, the negative results of the Timna excavations as far as the 'Mines' are concerned, are well corroborated by *Chronicles* I 22:3; 'And David prepared iron in abundance for the nails for the doors of the gates, and for the joinings; and brass in abundance without weight . . .' and *Chronicles* I 18:8; 'Likewise from Tibhath, and from Chun, cities of Hadarezer [King of Zobah], brought David very much brass, wherewith Solomon made the brazen sea, and the pillars and the vessel of brass . . .'

Amalekite settlements in the Negev

The absolute dates now available for the three kinds of pottery found in the temple help us to date many important sites in the Central Negev and the Arabah, in Edom and especially in 'the Land of Midian'. In the Central Negev, in the area south of the Jeruham valley and north of the Makhtesh Ramon, there existed numerous settlements based on dry farming in terraced wadi beds and the use of cisterns and run-off rain water for irrigation. These settlements and the casemate fortlets attached were generally identified as 'Israelite settlements' because Iron Age pottery was found amongst the ruins, including much of the unique primitive, hand-made, Negev-type pottery. Biblical associations rather than stratified archaeological evidence seem to have led to the dating of these settlements to the tenth to eighth centuries BC and their relation to the Judaean Kingdom. Continuous study of these settlements, their unique irrigation technology and pottery, had related the sedentary civilization of the Negev to the Amalekite tribes, mentioned in the Bible as the inhabitants of the Negev Mountains at least as early as the thirteenth century and down to the eighth century BC. Yet, the fact that in many of these settlements only Negev-type pottery was found, made the dating rather uncertain. Now, with the discovery of the Timna Temple, we have indisputable evidence that Negev-type pottery was made as early as the late fourteenth century BC and was in use for a very long period afterwards. The peculiar pottery-making tradition in the Negev, which could not possibly have originated in Judaean times and never occurs anywhere in Judah itself, would therefore exclude any possible identification of the Negev settlements as Israelite. Although not enough archaeological evidence exists so far for the accurate dating of these settlements, the Timna Temple finds strongly corroborate the view that many of the agricultural settlements and hill fortresses in the Central Negev predate the Israelite conquest of Palestine and already existed as fortified Amalekite villages at the time of the Exodus. It therefore seems plausible to conclude that some of the battles between the Israelite tribes on their way to the Promised Land and the Amalekites, their arch enemies, must have taken place around these settlements and fortresses. It appears also most likely that the destruction of many of the fortresses and settlements was actually caused by the continuous struggle carried on during most of the Kingdom of Israel between Amalekites and Israelites. Amalekites were still reported as settlers in the Negev Mountains as late as the time of Hezekiah, King of Judah. In *Chronicles* I 4:42–43 we find the latest date for Amalekite habitation in the Negev given as the end of the eighth century BC, whilst the Negev-type pottery found in the Timna Temple strongly suggests the existence of a sedentary civilization in the Central Negev at the end of the fourteenth and continuing well into the twelfth centuries BC. This latter conclusion is also based on the archaeological evidence of a direct connection between Late Bronze Age–Early Iron Age I Timna and the contemporary settlements in the Negev Mountains. Some small copper smelting sites with Negev pottery only were found along the paths leading from Timna through the southern Negev Mountains towards Makhtesh Ramon, as for instance Site 229. The copper ore for these smelters could only have come from Timna and the assumption is that it was carried north by workers of the Egyptian copper mines on their way home.

A date for Midian's towns

Other new aspects of chronology and of historical interpretation are provided by the absolute dating at Timna of the decorated Midianite pottery. This pottery had previously been found by the expedition in the smelting camps of the western Arabah and on the island of Jezirat Fara'un in the Red Sea. Prior to this some sherds of this ware had been found by N. Glueck during his survey of the eastern Arabah and Edom and called 'Edomite' ware. In 1935 Glueck dated this pottery correctly to the thirteenth to twelfth centuries BC. Yet, until it appeared in stratified and absolutely dated contexts in the Timna excavations, it could not be dated with any certainty and its origin also remained a matter of conjecture. Today the Timna types of Midianite ware are dated to the fourteenth to twelfth centuries BC and there is good archaeological evidence for its origin in north-west Arabia, in the area of Midian. Indeed, the survey report of Midian published in 1970 describes the site of a kiln at Qurayyah where this decorated pottery was actually produced and a Late Bronze Age date for at least some of it was correctly suggested by Peter Parr. At Qurayyah a whole sequence of decorated ware was found; some of it seems earlier than the Timna ware, other pieces seem later, but there can be no doubt that the Timna ware is fully represented at Qurayyah. Although the results of comparative analyses of the Qurayyah and Timna pottery are still awaited, it may confidently be said that the Timna pottery originates from Midian and provides the first certain absolute dates for the ancient town sites of Midian, where this pottery is to be found.

Midianite partnership at Timna

The appearance of pottery of clearly non-local origin does not of course imply automatically the appearance of foreign people and it is quite possible that it was commercially imported. However, this issue should be considered within the context of all objects found in the temple, including its architecture and furnishings and, not least, the overall picture of the Timna sites. It would be unreasonable to doubt that Midianites actually worked and worshipped in Timna. There was no Midianite pottery in the earliest, initial phase of the temple and it seems plausible to assume that at the very beginning of Egyptian copper mining in Timna the Midianites were not yet working there, while the Amalekites from the Negev were already present. Perhaps this fact explains the differences between Sites 30 and 34, located opposite the temple in Nahal Nehushtan and defended by a strong wall, and the other sites of Timna which had no defensive wall.

Whereas at Site 30 no Midianite ware was found (but there was an early stage of metallurgy, known to us from fifteenth to fourteenth century Egyptian smelters of Bir Nasib in Sinai) and very little Midianite pottery was located at Site 34, a large quantity of Midianite ware was found at the unwalled sites of Timna and also in the excavation of Site 2 where it appears from the very beginning and in all levels of the smelting camp. We may perhaps see here a repetition of the story of the Egyptian mines in Sinai where, after obvious conditions of enmity at the beginning of mining in the Wadi Maghara with defensive walls put up around the miners' camps and even a stronghold in the centre of the valley, the Egyptians, after initial setbacks, reached a peaceful working agreement with the local Semitic tribes, and defensive measures were no longer needed. In Timna, according to the evidence in the temple, the Midianites and the Amalekites, the indigenous inhabitants of the area, seem to have become some kind of 'partners' not only at work but also in the worship of Hathor.

The problem of the Kenites

We have no way of telling from the archaeological evidence found at Timna whether or not the Kenites, traditionally the ancient metalworkers of the Arabah, played an active role in the Timna copper works, but it seems likely. According to *Samuel* I 15:5–6 the Kenites were at times connected with the Amalekite Negev settlements and such a connection is also strongly suggested by the Negev ware at Timna. On the other hand, the Kenites are also identified as a clan or tribe of the Midianites and it may be the Kenites–Midianites of this Biblical tradition who made their appearance in Timna. It has already been suggested in the past that Jethro, the Kenite–Midianite father-in-law of Moses, had taught Moses to fashion the Nehushtan, the magic copper serpent, and the Midianite gilded copper snake in the shrine of the Timna Temple in its last, Midianite phase, seems to furnish a factual background for this tradition.

A Midianite tented shrine and the Tabernacle

The votive copper snake of Timna is only a part of the Midianite cult represented in the temple. Indeed, it is the first time that Midianite civilization and worship has come to light in the form of temple architecture and of a variety of votive objects and offerings. The study of the religious and cultural implications and connections of the Timna Temple finds is only at its beginning, but already at this stage of enquiry these implications seem important for the understanding of the formative phase of Israel. The last phase of the Hathor Temple of Timna, which seems to have been a tent-shrine, was a Midianite place of worship and this suggests a possible connection not only of the Midianite cult of the copper snake, found in this shrine, with the Nehushtan of the Exodus, but also with the actual tent-shrine of Israel's desert wanderings, the 'tent of meeting', the Tabernacle.

In the light of the Timna discoveries, it seems at least plausible to consider the tented-shrine, the Ohel Mo'ed, of Israel's nomadic desert faith to be somehow connected with the relationship between Moses and Jethro, who was not only a priest (*Exodus* 3 : 1) and advisor of Moses (*Exodus* 18 : 13–27) but also performed sacrifices and took part in a sacred meal 'before Yahweh' (*Exodus* 18 : 12). We recall here the view, voiced by some Biblical scholars, that the cult of Yahweh, at this stage intrinsically the invisible Yahweh who 'tented' among his people and whose proper dwelling was a tent, may have been of Kenite-Midianite origin. In this connection it seems relevant to recall also the obvious anti-Hathor practice of the builders of the tented shrine at Timna, though the defacing of Hathor and the destruction of the Egyptian monuments may also be otherwise explainable.

Timna and the Exodus

Whatever theological implications one attaches to the Midianite shrine, the discoveries at Timna provide a factual, cultural and historico-geographical background to the early desert part of the Exodus narrative. Yet, the presence here of Egyptians and Midianites in the thirteenth century, generally accepted as the period of the Exodus, is of course of great significance and raises many problems. There seems to be little doubt that the actual existence of a large scale Pharaonic industrial enterprise in the Arabah during the fourteenth to twelfth centuries BC will require reconsideration of the factual foundations of current Bible interpretations and historical concepts regarding the Exodus.

102–104 Three views of a small copper votive figurine (approx. twice actual size) from the copper hoard (Plate 95). It represents a bearded male, perhaps a fertility idol, of non-Egyptian character (Plate XVIII). The figurine was found in an 'as cast' state, with small parts of the mould still sticking to it, proving its local origin

105-107 Some examples of the curiously shaped stones and fossils, brought to the temple as votive gifts by local mining workers. The resemblance of some of these pieces to mother-and-child figurines obviously caught the imagination of the devotees, and suggested their suitability as gifts to the goddess of the miners

108 As well as the curiously shaped stones (*opposite*), the miners brought a varied selection of fossils and sea-shells to the temple. Among the examples here can be seen a fossil sea-urchin, various shells from the Red Sea, a piece of coral from the sea-bed, and some ore-nodules

109, 110 Places of worship at Site 2. *Above*, a High Place in Area F with the *bamah* on top and debris below (*Fig. 34*). *Below*, the five standing stones, *mazzeboth*, with a libation bowl in Area A (*Fig. 33*)

111, 112 A funerary shrine on top of 'King Solomon's Pillars' (Site 198). In a cave, created by an immense fallen stone (*above*) was found a *mazzebah*, carefully erected upon a small 'table' or altar (*below*)

113, 114 A small Semitic temple in Area A, Site 2. *Left*, Structure 2, its outline clearly visible among the debris before excavation. *Below*, the temple as revealed beneath the debris and after the removal of Structure 2. Its Semitic nature is undisputable on the evidence of the five standing stones (Plate 110), the central altar, offering bench and the attached priest's cell (*Fig. 33*)

115 Roman copper mining was extensive throughout the southern Arabah (*Fig. 64*) as well as in the Timna Valley. Wherever a cupriferous white sandstone appeared in the area the Romans worked it, apparently under the direction of the Third Legion *Cyrenaica*. Site 23 in the Timna Valley shows the typical Roman shaft-mining technology

116 Site 24a, in the northern Timna Valley, illustrates the Roman technique of copper mining. Deep galleries were cut into the white sandstone following the seam of rich copper nodules. Within the mountain galleries branch out and are often on several levels. The gallery walls still clearly show the marks of the Roman mining tools (Plate 118)

117 A Roman memorial inscription, in Greek characters, was found near the entrance to a tunnel near Site 23. It seems to commemorate several miners who died in the area, Aurelius, Eutuchis, Zanthius and Theon. Above the inscription a Roman eagle has been scratched onto the rock face

118 A view into one of the Roman mining galleries showing the marks of the metal tools used in cutting through the white sandstone

119 View of the Roman copper smelting site (No. 28) near Beer Ora. The extensive slag heaps visible prior to excavation were an obvious indication of the large scale copper smelting at the site, and led to the excavations in 1969 (*Fig. 65*)

120 Circles of tapped copper slag as found in heaps next to the furnaces at Beer Ora. Note the cast-in holes in the centre of the pieces, to facilitate their quick removal from the tapping pits

121 A strange structure composed of lines of slag pieces in Area D, Site 28. It is rectangular in shape with an apse on its east and south sides (*Fig. 71*). It has been suggested that it may be a place of worship, possibly a 'symbolic' early Christian church

122 Roman smelting furnace in Area F, Site 28, showing the smelting bowl and part of the slag pit in front of it (*Fig. 69*)

123 Partially excavated Roman smelting furnace: Area A, Site 28 (*Fig. 68*)

124 Roman crucible melting furnace: Area E, Site 28 (*Fig. 70*)

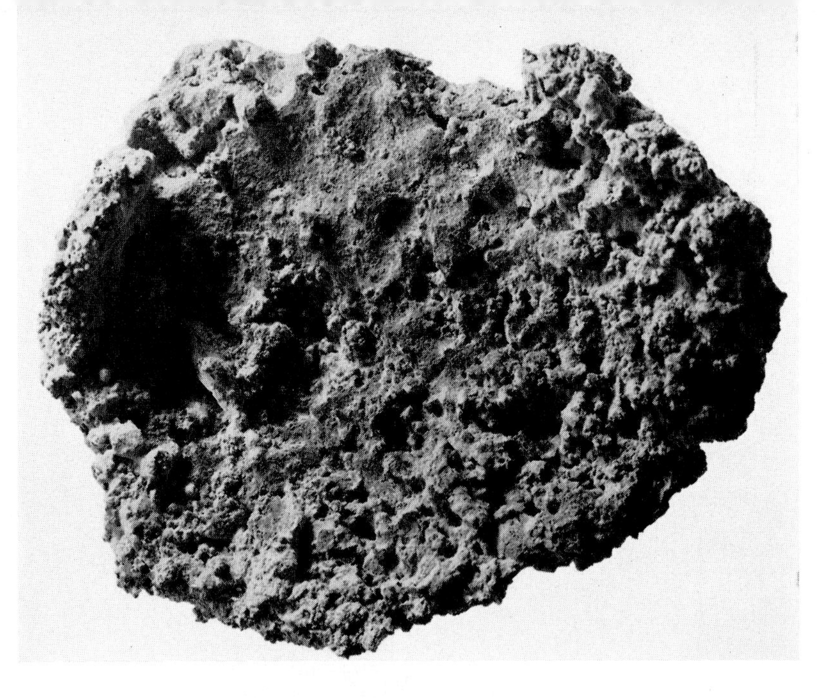

125 Convex lump of a smithing cinder from the bottom of a Mameluke furnace at Site 224 (*Fig. 76*)

126 Stores and workshops were erected at Beer Ora using large lumps of copper slag as building material (*Fig. 67*)

127 A Mameluke smithy (Site 224) on the Darb el-Hagg (*Fig. 75*). The furnace stands in the centre of the smithy (Plate 128). In the background lies the Mameluke workers' camp of the fourteenth century AD with its numerous stone-built structures

128 The smithing furnace in the Mameluke Site 224. On the left are the remains of the two circular chimneys and, to the right, the hard-burnt furnace bottom. Numerous examples of the iron products from this furnace were found on the site (*Fig. 77*). This furnace has now been removed from the site and reconstructed in the Museum Haaretz, Tel Aviv

Arabah = Atika ?

Although there is sufficient evidence in Egyptian sources for Ramesside military campaigns in the Negev, Edom and the Arabah, the Hathor Temple of Timna provides the first archaeological evidence for actual and lengthy Egyptian control of this area. We still lack reliable archaeological information on the copper mines and smelting camps of the eastern Arabah and Edom, but we may safely assume that the Egyptians also worked these mines and that the whole of the Arabah was controlled by the XIXth and XXth Dynasty Pharaohs.

A highly interesting record of large scale copper smelting undertakings exists from the reign of Ramesses III, though until now this detail was hardly noticed and the name of the mining area mentioned remained so far unidentified. In Papyrus Harris I (408) we read: 'I sent forth my messengers to the country of Atika, to the great copper mines which are in this place. Their galleys carried them, others on the land journey were upon their asses. Their mines were found abounding in copper; it was loaded by ten thousands into their galleys. They (it) were sent forward to Egypt and arrived safely. It was carried and made into a heap under the balcony, in many bars of copper, like hundred-thousands, being of the colour of gold of three times. I allowed all the people to see them, like wonders'. Atika was a copper-bearing region, accessible both by sea and land from Egypt and in the light of the evidence from Timna we propose to identify it with the Arabah, and particularly with Timna, the only area known to us so far which is rich in copper ores, is accessible both by sea and land from Egypt, and shows copper mines and smelters of the time of Ramesses III.

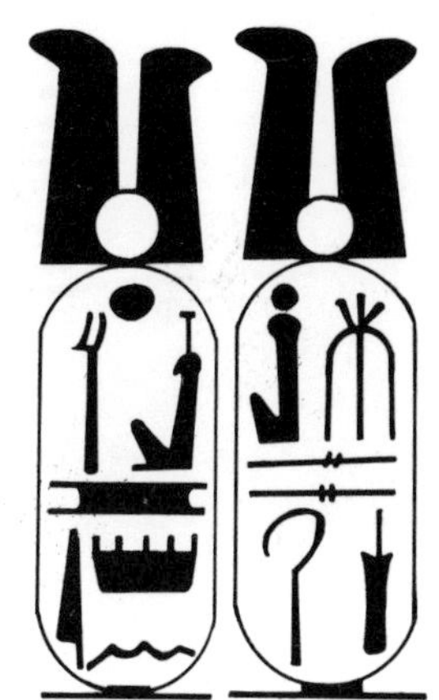

62 Monumental double cartouche of Ramesses III, 40 × 60 cm., found on the ancient road from Sinai to the Arabah copper mines

A discovery made in February, 1972, in the mountains of Elat provides additional evidence for our proposed indentification of 'Atika' with the Arabah. A monumental hieroglyphic inscription, carved into the rock-face at an ancient watering place north of Ras en-Naqb – now called Borot Roded (Site 582, GR 13798920 Isr. Grid) – was first noticed by a young schoolboy from Elat and copied and investigated by the author. It is a 40 × 60 cm. large double cartouche of Ramesses III, each cartouche is surmounted by a disc-and-feather crown. The inscription, apparently proclaiming the waterholes of Nahal Roded as the Pharaoh's property, is evidence for Ramesside control of the path which leads from Sinai through Nahal Roded into the Arabah. It may well be the road taken by the Egyptian mining expeditions' overland donkey caravans to 'Atika'. Two camping sites with fourteenth to twelfth century pottery, can now be interpreted as stations on this road.

Jezirat Fara'un – an Egyptian mining port?

Plate 63

The Timna discoveries may also help to throw light on the highly interesting remains on the island and in the bay of Jezirat Fara'un, located in the Gulf of Elat, some 4 km. south of Taba. This island was called variously 'Ile de Graye', 'el Qureiye', 'el Deir', 'el Kasr hadid', 'Emrag', and recently the 'Coral Island'. It was first described by E. Rüppel in 1829 and has since been repeatedly visited by travellers, geographers and archaeologists who dated the remains on the island as Byzantine, medieval Arabic or Crusader. Only C. L. Woolley and T. E. Lawrence noticed that 'all round the shore at sea level are to be seen the remains of a wall built of rough masonry about 4 feet thick, entirely destroyed down to the level of the beach'. Yet, Woolley and Lawrence refrained from dating this 'first wall' of the island.

Fig. 63

In 1956–57 the author investigated the island, together with A. Hashimshoni, and published a modern survey plan of most of its remains. The island is 320 m. long, its maximum width is 150 m. and the straits separating it from the mainland are only about 275 m. wide. Its most conspicuous ruins are the remains of the medieval castle, located on three steep granite hills. Detailed descriptions of these ruins and other features of the island have been published in

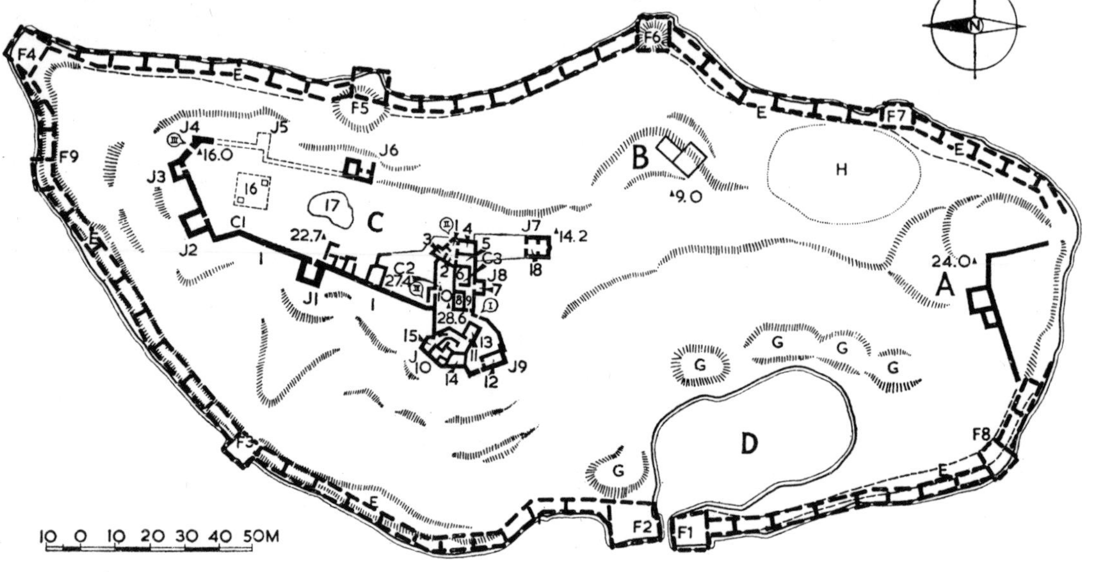

63 Plan of Jezirat Fara'un, the 'Coral Island' south of Elat. The plan of the island was prepared by A. Hashimshoni; the casemate walls around the island are a schematic reconstruction by the author

God's Wilderness, Discoveries in Sinai and in *Negev* and need not be repeated here. The remains relevant to this chapter are located on the shore-line around the island and were identified as casemate walls of cyclopean character. In this wall traces of defensive towers, projecting out into the sea, are discernible. There are also foundations of houses (Area H) at ground level on the island, and a small harbour with its entrance facing the mainland.

Plate 64

At the time of the first investigations on the island in 1956–57 we collected some pottery which was mainly Roman-Byzantine and medieval Arabic. There was also some rough and some decorated pottery which at first seemed chronologically unrelated, but in 1961, after the first Timna survey, it was identified as Early Iron Age I pottery. It was clearly the same Midianite (called at the time 'Edomite') and Negev-type ware which had been found in the Timna smelting camps. Some of the sherds, especially fragments of a cooking pot, could also belong to Iron Age II. In the light of these pottery dates proposed at the time, and in detail in *Negev*, the following working hypothesis on the history of the island was put forward: the earliest remains at the site, which consist of the casemate wall, the harbour wall and dwellings in Area H, and remains of a landing pier on the mainland opposite, date to the Early Iron Age I prior to the United Monarchy of Israel, with a possible use also in the tenth to eighth centuries BC. All other remains relate to the Nabataean, Byzantine and Mameluke occupation of the island. Furthermore, in view of these dates and the fact that the island is the only natural anchorage in the northern part of the Gulf of Elat-Aqaba, it seemed logical to look here not only for a harbour of the period before David's conquest of the area, but also for the port used by Israel's kings. In other words, it was proposed at the time to identify the island of Jezirat Fara'un with the Biblical harbour of Ezion Geber.

With the new evidence from the Timna Temple relating the copper mining activities of the Late Bronze Age–Early Iron Age I to the Ramesside pharaohs, and taking into consideration the Papyrus Harris I report of the existence of a regular shipping route to Atika and its identification with the Arabah mines, we now propose to identify early Jezirat Fara'un as a Pharaonic mining harbour. This Egyptian mining port would later be the obvious anchorage for King Solomon's Tarshish ships as it was the main and probably only safe port of the northern Red Sea during all subsequent ancient periods.

Since 1967 the expedition has re-investigated the island of Jezirat Fara'un and a large amount of pottery was collected there. These investigations and the finds on the island confirmed once more the dates previously proposed, including the existence of fourteenth to twelfth century BC Negev and Midianite ware. It became increasingly clear that only systematic excavation of a large part of the island would archaeologically justify the detailed dating of the individual structures on the island. More small finds were found, including medieval Arabic inscriptions and, on the western slope of hill A, remains of a small metallurgical installation and a quantity of fayalite slag, evidence for small-scale iron smelting activities on the island. This may explain one of the island's Arabic names – 'el Kasr hadid', the Iron Castle – though this may well be a poetic connotation of a highly romantic site.

Whilst this chapter was being written, Alexander Flinder, architect and undersea explorer, sent a short report on his investigations of Jezirat Fara'un which contains many details of considerable interest. With his kind permission it is included here:

'In the autumn of 1967 I joined Dr Elisha Linder and a small group of divers from the Undersea Exploration Society of Israel on a short reconnaissance of the undersea terrain around the island of Jezirat Fara'un in the Gulf of Eilat. In the following year I returned to lead a combined Anglo-Israeli group in a systematic search of the sea-bed and of the shores of the island, and the mainland.

'Although our undersea discoveries were rewarding – a large group of Byzantine period amphorae, and basalt grinding mills; it was as an architect that I was drawn to examining the intriguing remains of buildings and structures on the island itself. These have been described by Dr Beno Rothenberg in his books, *God's Wilderness* and *Negev* – the twelfth century Moslem palace on the highest of the island's three hills, the Byzantine buildings on the southern and lower hill, the remains of the casemate wall which circumscribes the perimeter of the island, and the small harbour nestling in the hollow between the two hills.

'I concentrated on the casemate wall and the harbour, and as my sketches and measurements developed I became increasingly aware of the subtle reasoning that went into the design of what must have been a splendid example of ancient marine-defence building. The wall comprises an outer skin facing the sea, formed of cyclopic fashioned stone blocks, the largest being nearly two metres long by one metre thick. There is then an inner skin formed of smaller blocks and about 50 cm. thick. This runs parallel with, and 2·30 metres away from, the outer skin, and this space is filled with concrete rubble. Thus the total thickness of the compound wall is nearly four metres. The casemate part of the wall is formed by cross walls coming out at right angles from the inner skin at regular intervals, and the 'rooms' formed by these and the innermost wall measure in excess of three metres by two metres each. The overall thickness of the wall with its inner casemate rooms is over six and a half metres. A truly mighty structure built to withstand the fury of the violent storms that can develop so rapidly in this short strip of the Great Rift Valley. The length of this wall is about 900 metres and is interrupted at more or less regular intervals along its perimeter by seven square towers which project out into the sea. It appears to be perforated in only two places:— in the north-west, where the narrow breach has an appearance of a slipway, and at the entrance to the harbour in the south-west. I was most fascinated by this harbour, because it is quite clear that its entrance and the casemate wall are integral and were constructed at the same time.

'Standing on the peak of the southern hill, one is able to look down on the island below and absorb the scope of this magnificent maritime installation – the thick casemate wall with its towers, curving round to embrace and enclose the small harbour; its one entrance leading from the narrow straits which separate the island from the mainland. In my view it is this slim stretch of sea that holds the key to the whole history of the development of Jezirat Fara'un; for here we have an anchorage, the best natural haven for ships in the northern half of the Gulf.

'I had occasion to return to the island recently in the middle of a violent storm. Hailstones spat down from a black sky and the sea was turbulant – except for the anchorage which was comparatively still especially close to the island where the surface was scarcely ruffled. Sir Richard Burton and Lt. Wellstead in the nineteenth century both testified to the effectiveness of this anchorage.

'The construction of a small protected harbour leading from the stillest part of the anchorage was a logical development; here then we have a natural anchorage a few miles from the northernmost tip of the gulf. An island beside the anchorage, protected by an impregnable wall – and harbouring within its defences yet a further shelter for boats. The seas around the island – a perfect moat! What a superb arrangement I thought! and yet something appeared to be missing – where and how did the island relate to the mainland. There must have existed a ferry, a means of crossing from the mainland to the island, and therefore somewhere on the beaches of the mainland we should find some evidence of this. We searched in vain, and had virtually given up hope, when one day a lucky combination of a very low tide and an exceptionally calm sea revealed to us a perfect stone jetty fifteen metres long by six metres wide. It was just where it should be – directly opposite and the shortest distance from the harbour entrance.

'The harbour itself is extremely interesting. I had initially considered it to be rock-cut, but I now suspect that this is not so. I am inclined to think that it was originally a small open bay with a sand beach. The harbour was formed by enclosing this bay with a breakwater built across its open side. A defensive casemate wall was built on the breakwater and a narrow entrance left at one end. The entrance appears to have been partially filled in with rubble during a later period but the south flank of the entrance is clearly indicated by a return wall of large stone blocks which itself forms the side of a square tower of similar proportion to the other towers. The north side of the entrance is, however, not a repetition of the south, for instead of the right-angled edge of a tower, we see what appears to be the foundation of a wall gently curving into the harbour. This wall commences at the area of exceptionally calm water that I have described previously. The design permitted boats to be manoeuvred gently into the harbour from the anchorage and vice-versa, and this delightfully thought out detail was supplemented by two stone built piers in the sea just in line with the entrance. These piers, sometimes known as 'dolphins', would assist boats coming in from the south and the open sea to make fast before being manhandled into the harbour. They are now only visible under the sea.

'Conclusive dating of these absorbing structures must be awaited but of one fact I am certain:—the men who conceived the defensive wall and harbour of Jezirat Fara'un were men of the sea, and the most skilled of masterbuilders.'

Alexander Flinder's report of his investigation on the island and in the sea around it added many important details and new maritime aspects to our previous picture of Jezirat Fara'un, the 'Island of the Pharaohs'. Unfortunately, the investigations of his team, underwater and also on land, seems to have been inconclusive as far as the date of the casemate wall and harbour is concerned and systematic excavation on the island is still required. Although we still believe that the cyclopean casemate wall along the shore line and the harbour belongs to the earliest constructional phase of the island's defences it is, of course, possible that the earliest period of occupation of Jezirat Fara'un, testified by pottery and other finds, could pre-date the earliest structures on the island. There was no need for such enormous defensive walls to use the bay and island of Jezirat Fara'un as a safe anchorage. This was guaranteed by their natural geographic features. There is also no need whatsoever to assume that an ancient harbour for occasional use, and as such we conceive the Pharaonic anchorage as well as Solomon's Ezion Geber, must have been a large artificial port installation. Unless the Egyptians, or perhaps the Kings of Israel and Judah afterwards, built the enormous walls of Jezirat Fara'un to gain thereby an impregnable fortress, in addition to a safe harbour, they could have done very well with an undefended anchorage.

VII

Roman Copper Works in the Arabah

The Roman copper mines

Fig. 64
Plate 115

Already in 1959, at the beginning of the Timna survey, Site 23 at the head of Nahal Timna presented a peculiar problem. The site, located at G.R. 14279111, is reached after a short walk through a narrow winding canyon, cut by the floods into the red Nubian sandstone. Before reaching this canyon, the formations along the wadi are all of the same red sandstone with, here and there, some alluvial conglomerate intrusions. Walking through the 30 m. long canyon one ascends into an area of white sandstone, in fact, right into an ancient copper mine, Site 23. Like all ancient mines in the Timna valley it is an area of low white sandstone formations containing copper ore nodules, but here we meet for the first time with real shaft mining. Whilst the white sandstone hills everywhere else showed evidence of 'open cast' mining, typical of the Late Bronze Age–Early Iron Age I, Site 23 showed many 1–1·5 m. high and less than 1 m. wide openings to horizontal shafts, often several metres deep, with low and narrow galleries radiating off. The walls of both shafts and galleries showed clear marks left by narrow metal chisels.

Plate 118

This obviously represented a different mining technology but, at first, no pottery was found anywhere near or in these mine-shafts to indicate their date. The Roman date of Site 23 was only established after very minute and insistent searches for pottery in all the shaft mining sites of the Arabah. In the course of these investigations a small cross was found carved on the wall near the entrance to one of the mining shafts at Site 23, strongly suggesting an early Christian period for this mine.

The Roman mine at Site 23 was clearly connected with Site 7, located at G.R. 14319111, on the other, lower side of the small red canyon mentioned above, still in the red Nubian sandstone area. Here, a tunnel about 20 m. long, 5 m. wide, and 3 m. high was washed out of a conglomerate sand and stone deposit that bars the way of the run-off floods rushing down from the Timna cliffs into Nahal Timna. Inside the tunnel some Roman pottery dating from the first to second centuries AD was found in several primitive enclosures

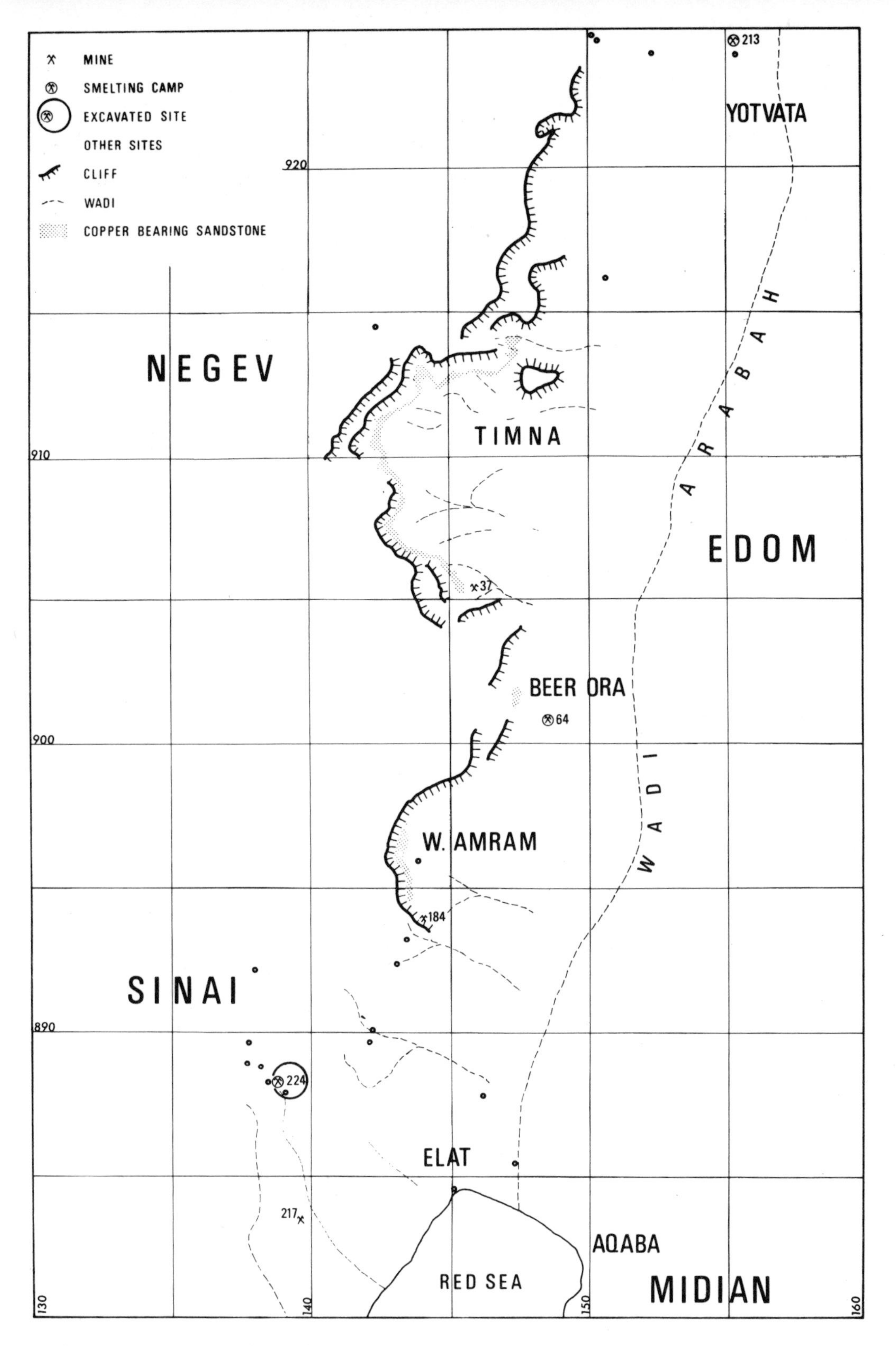

64 Roman sites in the Arabah and adjoining Negev mountains

Plate 117

of a rather temporary nature. Near the upper opening of the tunnel an inscription in Greek letters was carved into the wall, topped by a rock-carving of an eagle standing straddle-legged and with a laurel (?) wreath in its beak. Only part of this inscription is readable. It starts with 'In memory' (ΜΝΗΣΘΗ) and mentions further on the names Aurelius (ΑΡΗΛΙΟΣ), Eutuchis (ΕΥΤΥΧΙΣ), Zanthius (ΞΑΝΘΙΟΣ) and Theon (ΘΕΩΝ). A cross is scratched into the wall next to the inscription. The names were not carved into the wall at the same time, nor by the same hand, but must have been added every time a man died and was buried nearby. It is clearly a memorial inscription and seems to be related to several tumuli found just outside the upper entrance of the tunnel. The Roman miners, perhaps Christian slaves or convicts, working at Site 23 must have had their living quarters inside the tunnel of Site 7, which was a very convenient camping site.

Plate 116

Other Roman shaft mines, similar to Site 23, were located in the area of Sites 24A, (G.R. 14459119), 25A (G.R. 14479124), and 212 (G.R. 14329115) in the Timna valley. Here Roman pottery of the first two centuries AD was collected next to the actual mining sites but no Roman copper smelting remains were found.

Timna was not the only Roman mining area in the Arabah. Several extensive mining sites are known to have existed on the east side of the Arabah, from the area of Feinan in the north down to Wadi Abu Khusheiba, south-west of Petra, and Roman and Nabataean pottery and coins were reported from there. At least one of these mining sites was mentioned by Eusebius though he refers to late Roman times. The Timna and Nahal Amram sites are as yet the earliest Romano-Christian mining sites identified. Although Feinan seems to have been the largest mining area in the Arabah operated by Christian convicts, evidence for the earliest Christian period is still awaited from there.

Plate IV

In the western Arabah further Roman shaft mines were located in Nahal Amran (Sites 38 and 72) and Nahal Shechoret (Site 184). South of Elat several more Roman mines were found in Nahal Tuweiba and Nahal Murah (Sites 301 and 319 on the Sinai Survey map) and also further south. All of these mines, dated by pottery to the Roman period, show the same horizontal shaft-and-gallery mining techniques.

Roman copper smelting sites

Slag heaps, remains of furnaces and the ruins of workshops of various size are found at a number of sites in the western Arabah, from Yotvata down to Elat. Some of these sites are mainly copper smelting installations, others are habitation sites or military establishments with a small metallurgical workshop attached. Pottery, found in quantity on all Roman and Nabataean sites of metallurgical activities in the western Arabah, ranges in date from the early first century BC to late Byzantine times. This wide span of time during which several fundamental historical changes took place in the southern Arabah poses serious problems of correlation between various sites of different natures. It would take too much space to go into the details of the various sites and their problems and therefore a short description of only the historically and metallurgically more significant group of sites is given. Except for Site 28, at Beer Ora, none of these sites are as yet excavated and the facts reported here are the results of surface finds and analytical work in the laboratory.

Starting with the Late Roman–Early Byzantine period, the third to fourth centuries AD, there is a renewed, though very sparse and mainly military, occupation of the southern Arabah, perhaps as a result of Diocletian's reform of the military organization of the Imperium Romanum. A small fortlet, Site 11 (G.R. 15439218), was erected near the rich water sources of Yotvata, guarding the Roman road from the Negev Mountains to Roman Aila, on the shores of the Red Sea. A small slag pile near this stronghold indicates small scale copper smelting, presumably by the inhabitants of the fort. There are in fact no remains in the south-western Arabah of any large scale metallurgical activities during the Late Roman and Byzantine periods. The only site of this period of any metallurgical significance is to be found further south along the Roman road to Aila. Here, at G.R. 14758895, Site 4 consists of a heap of perhaps 50 tons of copper smelting slag and traces of a smelting installation. On a low hillock next to the slag heap are the remains of a small building and the pottery found here correlates this site with the fortlet, Site 11.

During the earlier Roman period, in the second century AD, a unique military establishment, Site 46 (1–7) was installed between the low hills flanking the road from Elat northward to Kibbutz Elot. This consisted of several groups of structures of obviously military character, resembling modern temporary army camps. Lined up between the hills were the ruins of what must have been mostly living quarters for men and animals. Not enough debris was found around the remains to have provided more than foundations or fences for temporary accommodation, probably in tents or huts. Besides early Roman pottery, some late Nabataean painted ware and also seventh–ninth century AD pottery, together with several medieval Mameluke sherds, was found, providing evidence for secondary use of these camp sites in later periods. At several of these groups of structures small quantities of copper slag indicate some minor copper smelting activities. The slag is of very good quality and the people who produced it must have known a lot about the smelting of copper. However, Site 46 (1–7) was no industrial establishment and no actual smelting installations were found by the expedition. It may be assumed that the inhabitants of Site 46 were connected with the mining and smelting establishments along the southern Arabah, either as actual workers or, more likely, as guards or work supervisors.

Plate 119

The main and only large-scale early Roman copper production plant in the area is Site 28 (G.R. 14829032), located near Beer Ora, just south of the Timna valley. This was first investigated in 1960 and partly excavated in 1969.

Excavations at Beer Ora

Fig. 65

Plate 120

Site 28 is located in a small valley, about one km. from the well called to-day Beer Ora, and consists of two large and several smaller slag heaps. At the time of the first examination of the site it was immediately noticed that most of the slag seemed to have been broken off large circular slag plates, 60–80 cm. in diameter. A number of complete slag circles were also found, most of them having a small hole in their centre. Whilst piling the slag on to the large heaps the workers 'built' high enclosures, perhaps as protection against the strong north wind or as some sort of defence. At the time it was not possible to be sure whether this had been done intentionally and contemporary with the smelting activities at the site. However, during the subsequent excavations it became evident that the Beer Ora smelters of the Roman period regularly used slag as building material.

During the surface survey little of the small, porous slag, so common in Site 2, was found. The slag was of the heavy, very dark, slag-circle type, though not all circles had holes in their centres, and many pieces were simply full round slag plates, often much larger than the circles with holes. At the time of the preliminary survey no explanation was found for the absence of the small slag type and the various forms of the circular slag. On the slopes north-east of the big slag concentration, numerous complete circular slag plates were formed into ring-like enclosures about 4 m. in diameter, presumed to be the actual sites of smelting (Area G on the excavation plan). This assumption was disproved by the 1969 excavations. Amongst

65 Site 28: Location plan of the excavations at the Roman smelting camp near Beer Ora

the slag lumps of charred and slagged clay were collected. Some of these were fragments of furnace lining but others were parts of tuyères, originally of tubular shape, used to ventilate the smelting furnaces. The tubes must have been 10–15 cm. in diameter, but their length is unknown and no slagged tuyère ends, so conspicuous on all the Ramesside smelting sites, were found at Beer Ora. One particularly interesting find at Site 28 was a large lump of metallic copper of circular shape still attached to a piece of slag.

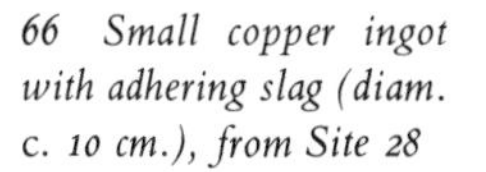

66 Small copper ingot with adhering slag (diam. c. 10 cm.), from Site 28

There were no stone buildings of any kind in the valley of the slag heaps and very few stones were visible among the slag; in fact there were no indications whatsoever of any stone-built smelting installations or furnaces. Some 300 m. north of the valley of the slag heaps, below a low hill rising in the estuary of Nahal Raham, a long line of ruined houses, each measuring about 5 × 2 m., was found. A quantity of Roman sherds and part of an oil lamp were collected among the debris as well as several simple stone implements, including a basalt mortar. Several small, stone-built, hearths were also found and it was assumed that the inhabitants of these houses (Site 28A) were the workers employed at Site 28, as no other habitation site was found in the vicinity.

Site 28 was excavated in 1969. The prime objective of the excavation was to find and unearth the actual smelting furnaces as well as enough archaeological data for the detailed metallurgical reconstruction of the smelting processes and the organization of the Roman copper industry in the Arabah. A 20 × 20 m. grid was laid over the whole site, measuring altogether 140 × 120 m., and excavations began simultaneously in eight different areas (A–H). Area C was abandoned after two days work as being utterly sterile.

AREA G
Fig. 67

Plate 126

In Area G, north-east of the main slag heaps, the above mentioned ring-like slag concentrations turned out to be two semi-attached structures. The main building was approximately 2 m. square, with an addition beside its entrance. This building showed a peculiar building technology: at first a square pit had been cut into the slope, down to a levelled floor of bed-rock, then large complete slag plates and circles were placed vertically against the newly cut surfaces, to keep out the sand and gravel. On the east and deepest side, the slag was standing on a foundation of stones and slag pieces, in order to achieve an even height all round. On top of this slag lining a wall was erected by piling heavy slag circles on top of each other, partly resting on the rim of the building pit, partly overlapping the standing slag-circle lining, to hold them firmly in place. The lower part of this building, including the slag lining, was found completely intact but the upper wall had almost entirely collapsed. On the floor a quantity of copper ore was found, indicating the building's function as a store house for raw materials.

The addition beside the entrance was a fireplace, fenced in by standing slag circles and found full of ashes, charcoal and pottery, including fragments of Roman cooking pots. This must have been the smelters' kitchen. Attached to the square building by a narrow, slag-lined passage, was a round structure which was simply a pit dug into the sloping ground and levelled off. It had no vertical slag lining but a wall had been built around the rim of the pit to a height of about 80 cm. This had almost entirely collapsed into the structure, apparently because of its great weight. A thick deposit of ash mixed with tiny bits of slag, also some Roman pottery and a copper awl were found on the natural rock floor.

Three metres south-west of these structures another round structure was excavated. Here vertical slag lining was found and the slag wall, resting on the edge of the construction pit, must have been more than a metre high. Pottery, as well as a copper piece, was found on a floor covered by a 10 cm. thick ashy deposit. The structure had been abandoned whilst still complete and started to fill in when the walls collapsed into the building and filled it almost completely with a solid mass of slag. It may have served also for habitation.

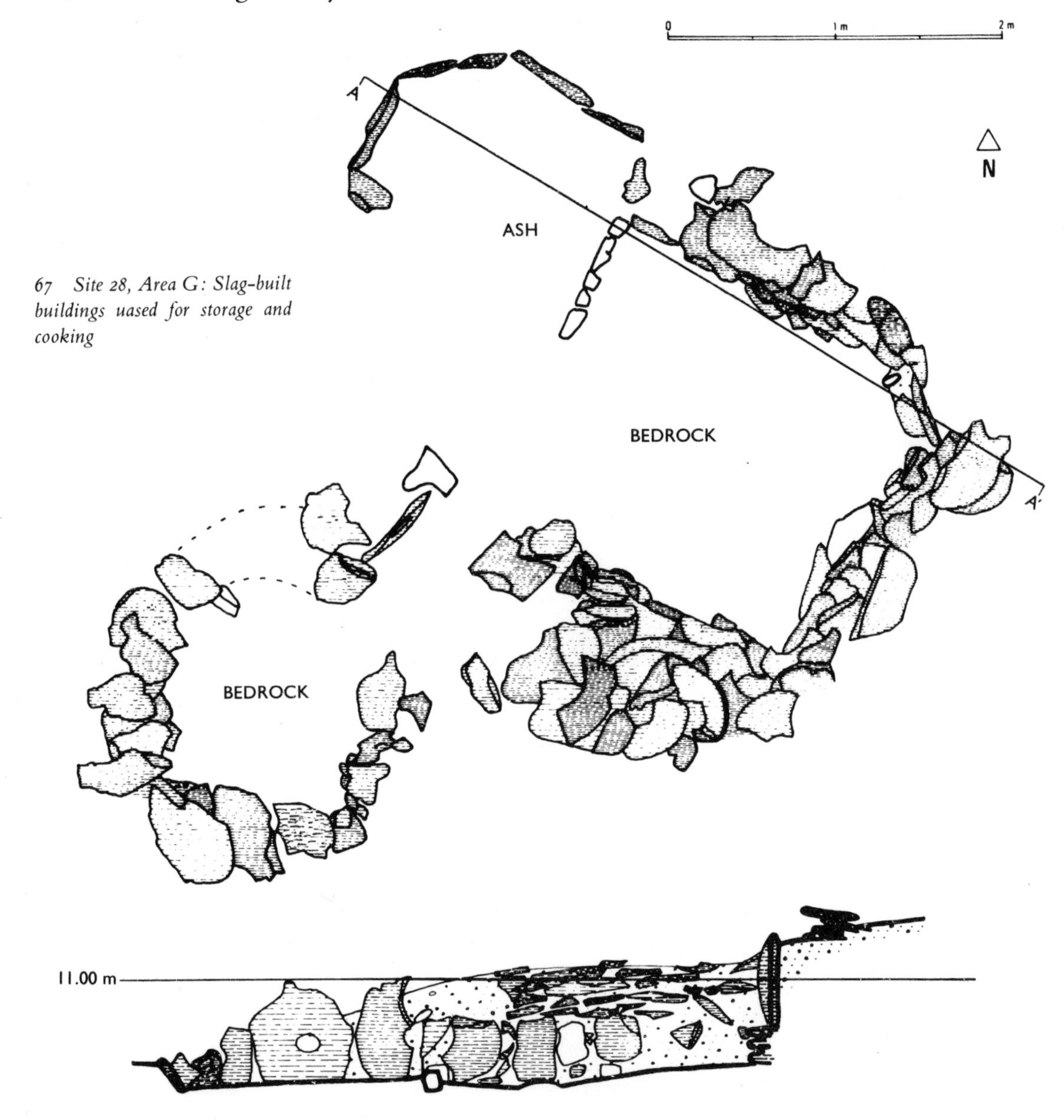

67 Site 28, Area G: Slag-built buildings uased for storage and cooking

AREA A – SMELTING FURNACE I

Fig. 68
Plate 123

A slight discoloration of the ground and a vaguely visible semi-circular accumulation of stones and slag next to the southern large slag heap, suggested a possible smelting furnace. After removing only several centimetres of the recent surface and some surface debris an old hard surface was reached and a semi-circular stone structure began to appear. It soon became evident that this was an almost complete Roman smelting furnace, Furnace I. Its smelting bowl was a pit, 20–30 cm. deep and 60–65 cm. in diameter, dug into the slightly sloping old surface. A row of small stones was placed in a semi-circle around its upper rim with an opening on its south side. Several moderately sized stones were found loosely surrounding the smelting furnace but not forming any real structure. These stones could have been so placed as support for the part of the furnace wall that rose above the level of the old surface. This uppermost part of the furnace, some of it found *in situ*, was built of smaller stones, and was clay-lined; it protruded 20–30 cm. above ground. The original furnace was probably 70 cm. high. The inside of the smelting bowl had a clay lining, 10–30 mm. thick, which in turn had a thin layer of slag adhering to it. The furnace bottom was also clay-lined with a thin layer of grey-burned clay dust on top. A shallow hollow, 1·7 m. in diameter, was dug next to the smelting bowl, reaching its maximum depth immediately below the bottom level of the slightly higher smelting bowl. This was the tapping installation with the actual slag pit next to the front wall of the furnace. A thick layer of charred material, charcoal and slag, was found in the slag pit. Its unlined bottom was of grey-burned soil on top of red-burned gravel.

The original front wall of the furnace had been removed during an earlier smelting process but was found closed again for an additional smelt by a compact mass of small stones and sand, with a thick layer of clay lining on the inside. However, this reconditioned furnace was never used. The whole furnace structure was found somewhat 'out of shape', perhaps as a result of an earth tremor. The bottom of the smelting bowl was no longer horizontal but sloped sharply down on one side and the reclosed front wall had been pushed into the bowl's interior. Debris fell into the bowl from above and added to the confusion.

No tuyère holes or any ventilation appliances were found and it was assumed that tuyères, many fragments of which had been found during the preliminary survey, were introduced into the smelting bowl through the front wall.

AREA F – SMELTING FURNACE IV

A second, similar, smelting furnace was unearthed in Area F. This furnace was completely excavated and almost completely destroyed

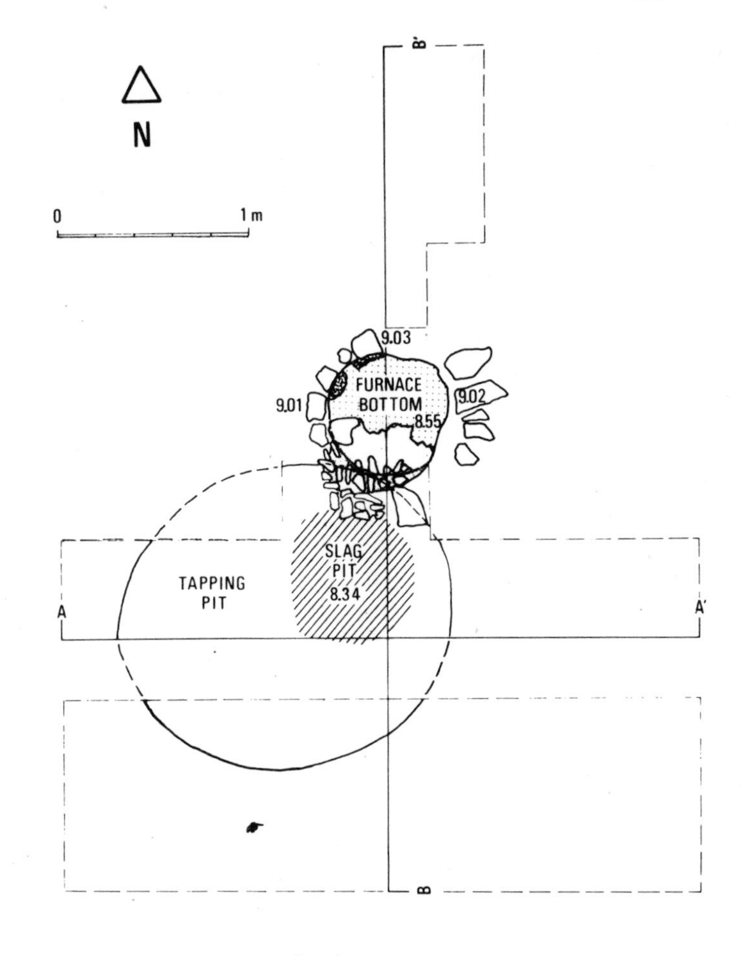

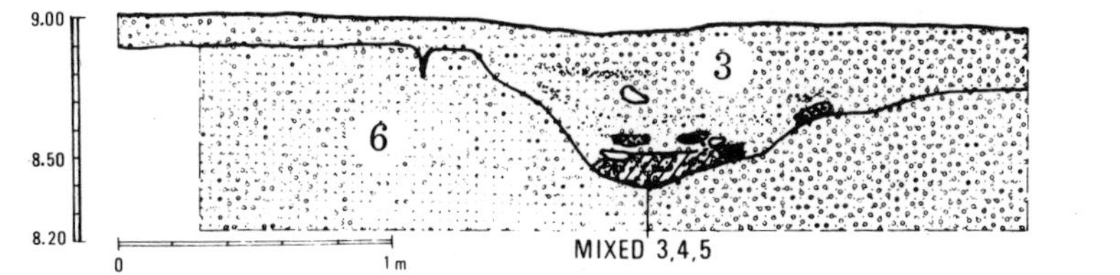

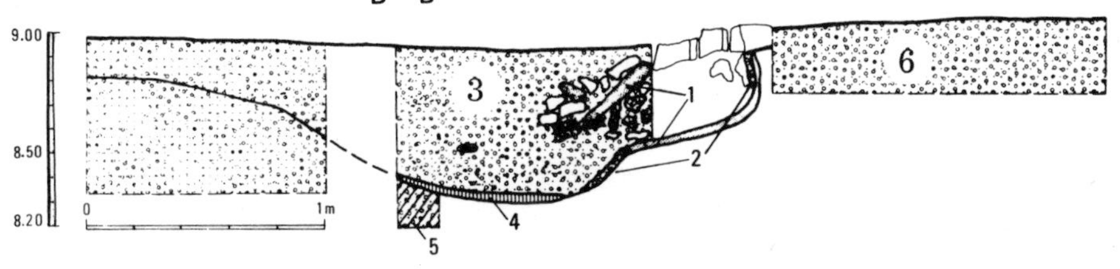

68 Site 28, Area A: Plan and section of Roman copper smelting furnace (Furnace I). Section details 1 Clay lining; 2 Slag; 3 Grey fill with metallurgical waste; 4 Ash tapping-pit bottom; 5 Hard burnt soil; 6 Undisturbed soil

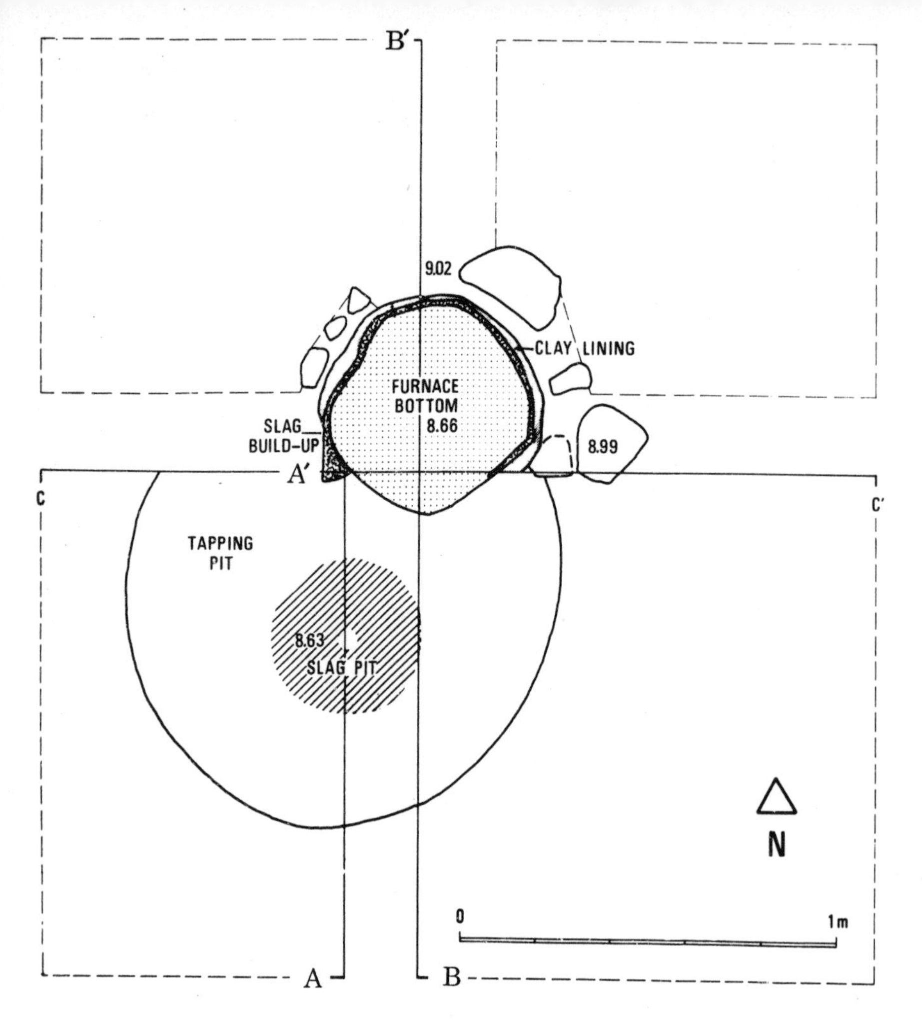

69 Site 28, Area F: Roman smelting furnace. The plan shows hard core in the centre of the tapping pit for the cast-in hole in the slag circles. Section A–A shows the tapping pit; section B–B and C–C show the furnace and tapping pit, both within the large pit dug for the furnace installation. 1 Clay lining on furnace bottom; 2 Slag; 3 Fill and metallurgical waste; 4 Ash; 5 Hard burnt; 6 Undisturbed soil

70 (Opposite): Site 28, Area E: Crucible melting furnace. The section shows the remains of an earlier crucible furnace beneath unburnt clay fill in front of the furnace. 1 Clay lining; 2, 3 Grey fill; 4 Wood ash; 5 Burnt soil; 6 Undisturbed soil; 7 Unburnt red clay fill ▶

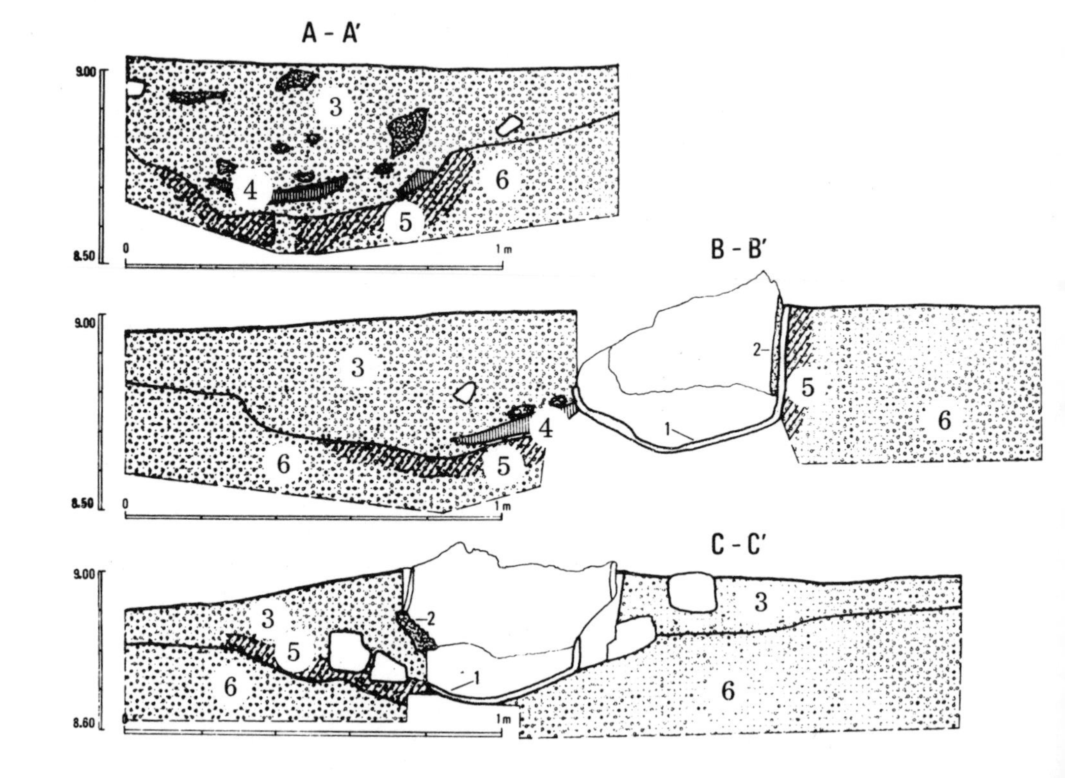

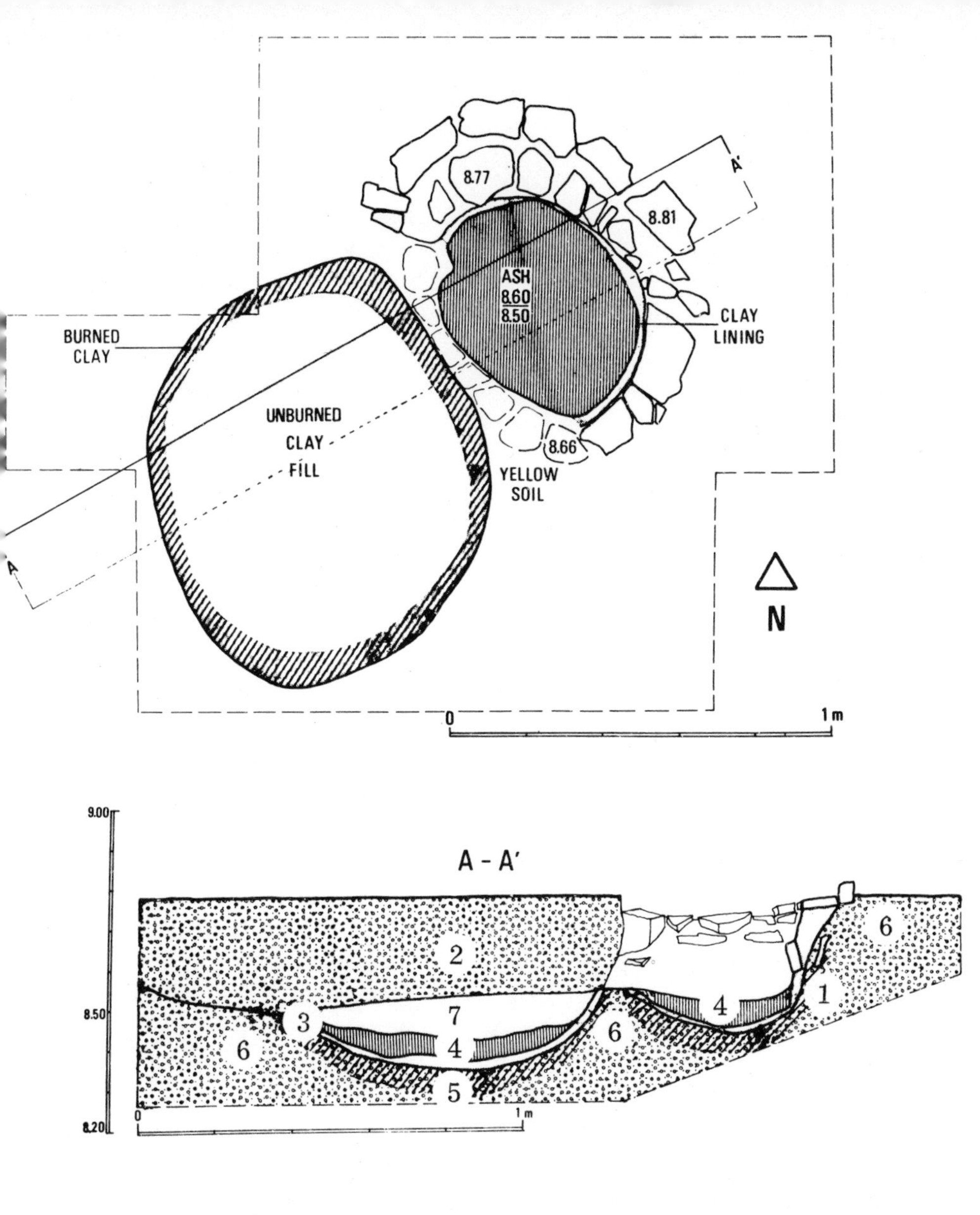

in the process to allow proper investigation of all its details. Numerous sections were dug, investigated and recorded. The most important detail of the furnace was the small, round and yellow piece of uncharred surface right in the centre of the slag pit. This spot must have been protected from the intense heat of the slag by some covering material. The assumption is that the original slag pit was dug in the form of a wide ring, with a hard core left in the centre for the casting-in of the centre-hole of the large slag circles.

Plate 122; *Fig. 69*

AREA E — CRUCIBLE MELTING FURNACE III

A small isolated pile of slag which included a large quantity of big, unburned charcoal pieces first drew attention to this spot. This seemed to be rough, porous furnace slag and a smelting furnace was therefore suspected in the immediate vicinity. A 4 m. square, Area E, was then excavated. About 25 cm. below the present surface appeared a shallow, oval-shaped hearth, 60 × 45 cm. and 30 cm. deep, which was found full of black-grey wood ash. A double row of small, flat stones must have enclosed the circumference of the pit, plastered with red non-calcarous clay. This hearth was certainly used for crucible melting of copper, as was proved by the charcoal and wood ash, numerous copper pieces and a quantity of melting slag found in Area F.

Plate 124; *Fig. 70*

A curious problem presented itself right next to this hearth. Whilst clearing the present surface of the recent upper sand and debris layer in order to find the old working floor, a completely flat, oval surface of red hard clay was uncovered, measuring 100 × 85 cm. Around the edge of this surface was a 5–10 cm. wide line of burnt clay. At first this surface was interpreted as a plastered working floor belonging to the hearth but this conclusion did not explain the line of burnt clay. A trial trench dug right through the hearth and the unburnt clay surface solved this problem. Before the hearth was constructed, an earlier, similar, crucible melting furnace had operated here. For some reason the Roman metalworkers decided to build a new furnace, instead of utilizing the old installation. They removed all the stones, to be subsequently re-used, and filled the remaining hearth, still full of charcoal, ashes, and bits of melting slag, with the same red clay used for the furnace lining, smoothing the surface on top into a clean hard floor. Why the Roman coppersmith went to all this trouble of filling in the shallow pit, and using good clay for this purpose, remains an intriguing question.

The discovery of a melting furnace, where copper was melted in crucibles prior to being cast, in addition to the two copper smelting furnaces, where metallic copper was produced from copper ores, completed the main objective of the excavations at Site 28.

AREA B

A segment of one of the large slag piles was completely removed, and the resulting sections enabled a study to be made of the inside of a slag heap, undisturbed for almost 2000 years. The uppermost slag layer was found to consist of clean, black slag but the almost solid mass of slag circles and plates underneath and down to the bottom was mixed with fine, yellow, wind-borne sand, which must have slowly infiltrated from the top and settled in the lower parts of the slag pile. This picture was quite interesting in comparison to the similar conditions found in the stone heap in Area F of Site 2 at Timna. Beneath the excavated part of the slag pile a thick and hard mass of charcoal dust was found and it seems that charcoal manufacturing was first practised at this spot. Many complete slag plates and circles, mostly with a hole in the centre, were found in the slag heap, measuring from 52–70 cm. in diameter and up to 7 cm. thick.

AREA D – AN ENIGMATIC SLAG STRUCTURE

Plate 121; *Fig. 71*

Lines of large slag pieces, sometimes double, were found protruding from the ground, forming a rectangular ground plan, 7·5 × 5·5 m., with an 'apse' at its south side. The whole area of this unique structure was excavated in order to find a possible floor and to try and find an explanation for its purpose. The slag lines had been intentionally constructed because the slag was inserted into deeply excavated, narrow, triangular foundation trenches and the 'apse' was constructed with particular care. The mode of construction of the slag lines proves that they could not have been intended to be more than an indication of walls, a marking off of a plot by means of a symbolic fence. On the floor inside the structure, 10 cm. below the present surface, some Roman pottery and traces of fireplaces were found, but nothing to show the function of the structure.

As a working hypothesis this segregated plot may be interpreted as a place of worship, possibly even a 'symbolic' early Christian church. The altar may have stood in the extreme east corner, which

71 Site 28, Area D: 'Structure' formed by slag pieces inserted into the ground, perhaps a 'symbolic' place of worship

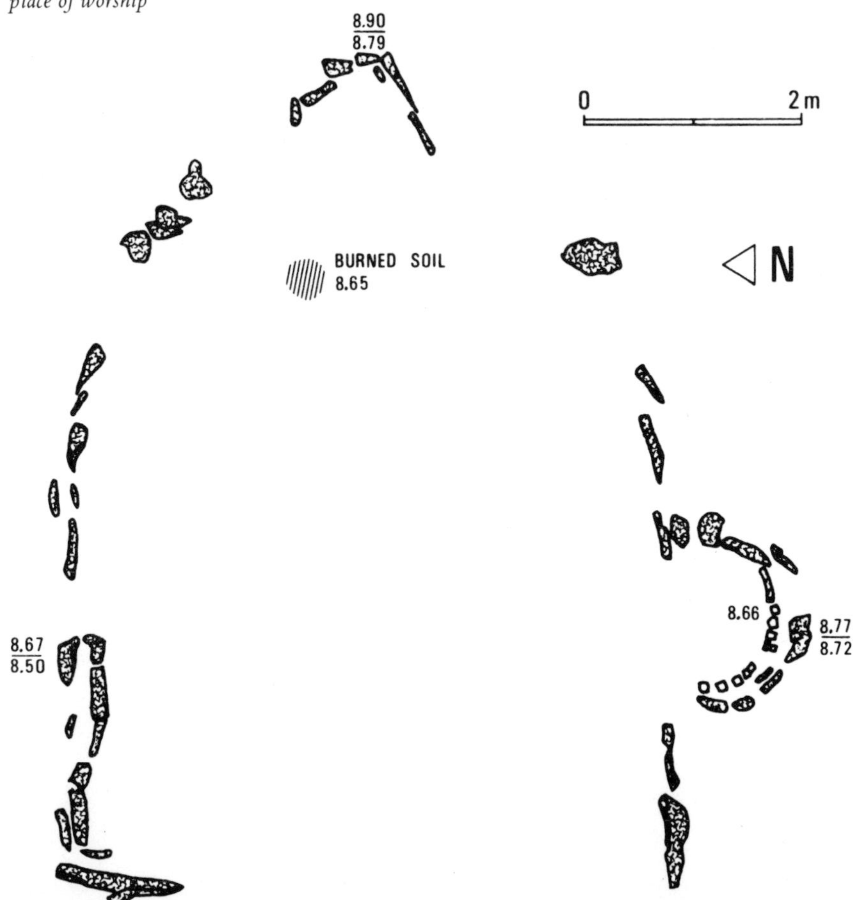

actually forms an addition apsidal recess, or in the 'apse' itself. This hypothesis derives from observations in Sinai of Beduin 'mosques', which are in fact only segregated plots, often marked out simply by a line of stones or sea shells, with an indicated 'apse' as the mikhrab. This Beduin tradition goes back to the earliest mosques which were at first only open spaces of assembly, often with some simple kind of rather flimsy demarcation.

The date of the Roman copper industry

Fig. 72

The pottery found in the excavation and preliminary investigations of Site 28 dated to the second century AD. It is only during this period that a fairly large copper industry existed in the south-western Arabah. From the Byzantine period to the Early Arabic period and, in fact, up to medieval times, all copper-smelting activities in the area covered by the expedition were of minor local importance only and were never on a higher level than 'home industries'. According to the literary sources and some archaeological evidence, the situation seems to have been quite different on the east side of the Arabah, especially at Feinan, where large scale Byzantine copper mines are reported. There the Nabataeans also seem to have been active in the copper mines, as in previous centuries. The Sinai survey produced a very similar picture of Nabataean to Byzantine copper mining activities, and it was therefore rather interesting to note that not one single Nabataean sherd was found in the large copper-smelting Site 28. The question arose as to whether the non-appearance of Nabataean pottery here and at the related copper mining sites of Timna and Nahal Amram, could help to date more accurately and perhaps identify the originators of the early Roman copper industry. Consequently, the date of the Roman copper industry in the western Arabah may be suggested as occurring after

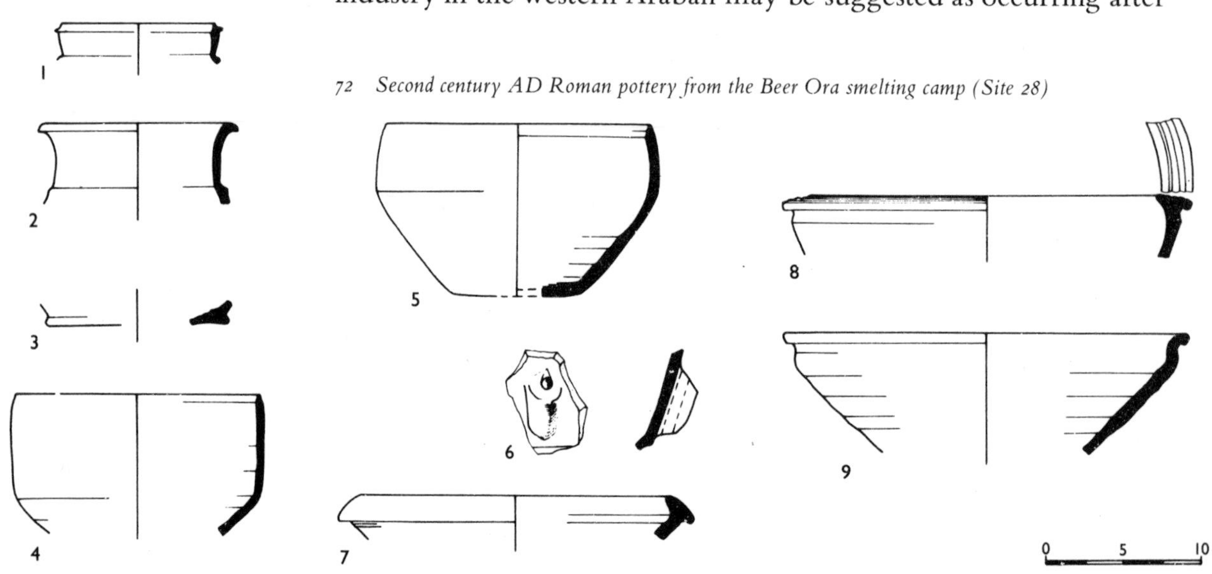

72 Second century AD Roman pottery from the Beer Ora smelting camp (Site 28)

the take-over of Nabataea by the Romans and to see it as part of the extensive and well-organized activities of the Third Legion *Cyrenaica*, stationed in the area at this time. Supporting evidence for this latter proposal comes from a Latin inscription discovered in 1933 by F. Frank in Nahal Tuweiba, reading, according to A. Alt: 'Tarry traveller, Here laboured and hence departed gladly T. Atilius Turbon, of the III Cyrenaic legion, of the century of Antonius Valens'.

73 Large rock-engraving found in the Arabah, perhaps late Roman period

In 1967 the Arabah Expedition investigated Nahal Tuweiba and prepared a new copy and photographs of Frank's inscription. This resulted in an improved reading and a different explanation of the nature of the inscription by S. Applebaum and E. Colman. It is now suggested that the inscription is an epitaph, and not a joke as in Alt's reading, and reads: 'Tarry, wayfarer, Here worked and died T. Atilius Turbon of the 3rd Legion Cyrenaica of the Century of Antonius. Farewell'.

At the time of its first publication in 1934/5 by F. Frank and A. Alt, its location in the rather isolated, and reportedly empty, Nahal Tuweiba at G.R. 13658777, near the shores of the Red Sea, was quite surprising. The location of this inscription is now also explainable because it is found on a huge boulder immediately below an ancient copper-mine and the men of the Third Legion *Cyrenaica* seem to have operated it. Furthermore, the Sinai expedition explored the whole length of Nahal Tuweiba and a large number of rock-drawings and many ancient remains were discovered there, dating to the fourth millennium BC, as well as to the Nabataean, early Roman, and Byzantine to early Arabic periods. As a result of the survey it also became evident that Nahal Tuweiba was the major ancient ascent from the Red Sea to the heights of the Sinai plateau and the Roman road Aila–Pharan, known from the Tabula Peutingeriana ('Haila–Phara', IX, 3–4), must have passed here as well. As the Third Legion *Cyreniaica* left Egypt *c.* AD 120–27 and sometime later, perhaps during the reign of Hadrian, was garrisoned at Bostra, the new capital of the new Roman province of Arabia, it is certain that the southern Arabah and its copper mines fell also within the area of its activities. It may therefore be assumed that the Third Legion *Cyrenaica*, which was already renowned for its earlier mining enterprises in Upper Egypt, started and operated the Roman copper works in the Arabah.

VIII

Medieval Arab Metalworkers

Fig. 74

The early Roman copper industry of the second century AD was the last proper industrial undertaking in the southern Arabah before the modern Timna Mines Ltd renewed the ancient mining tradition of the area. In a number of Late Bronze–Early Iron Age as well as Byzantine sites, *e.g.* Site 11e near Yotvata or Site 46/2 near Elat, some Arabic pottery was found, testifying to a secondary habitation in earlier buildings or just temporary use of convenient camping sites. The same applies to the metallurgical sites of the area. There is not one purely Arabic mining or smelting site anywhere, though in several early copper mines as well as in early smelting camps some early medieval Arabic sherds were found. Site 37, at Timna, and Site 184, south of Nahal Amram, can serve as good examples of secondary Arabic exploitation of earlier copper mines. This, of course, fits well into the general archaeological and historical picture of the area which shows an almost complete lapse of sedentary habitation in the southern Arabah since the Arab conquest. If, nevertheless, some evidence of Arabic metalworking exists in the area, it is for the secondary exploitation of earlier slag heaps or mining waste, probably by nomadic or semi-nomadic visitors to the earlier sites. There is, however, evidence for small-scale but professional metallurgical activities during the Mameluke period, mainly in the form of small slag heaps and metallurgical installations along the Darb el-Hagg.

The Darb el-Hagg and Site 224

The Darb el-Hagg, the Moslem Pilgrims' Road from Egypt and Sinai to north-west Arabia, and thence south towards Mecca, reaches the mountain pass above Elat at Ras en Naqb. From here the winding descent through the steep mountain sides is mainly the great construction work of the Fatimid Caliph Ahmed Ibn Tulun (AD 868–884), improved on and continuously repaired by later Mameluke Sultans. A memorial inscription, mentioned by Jarvis as having been found along this road, claims that the last of the Mameluke Sultans, El Ashraf Khansuh el Ghuri (1500–1516), ordered 'the

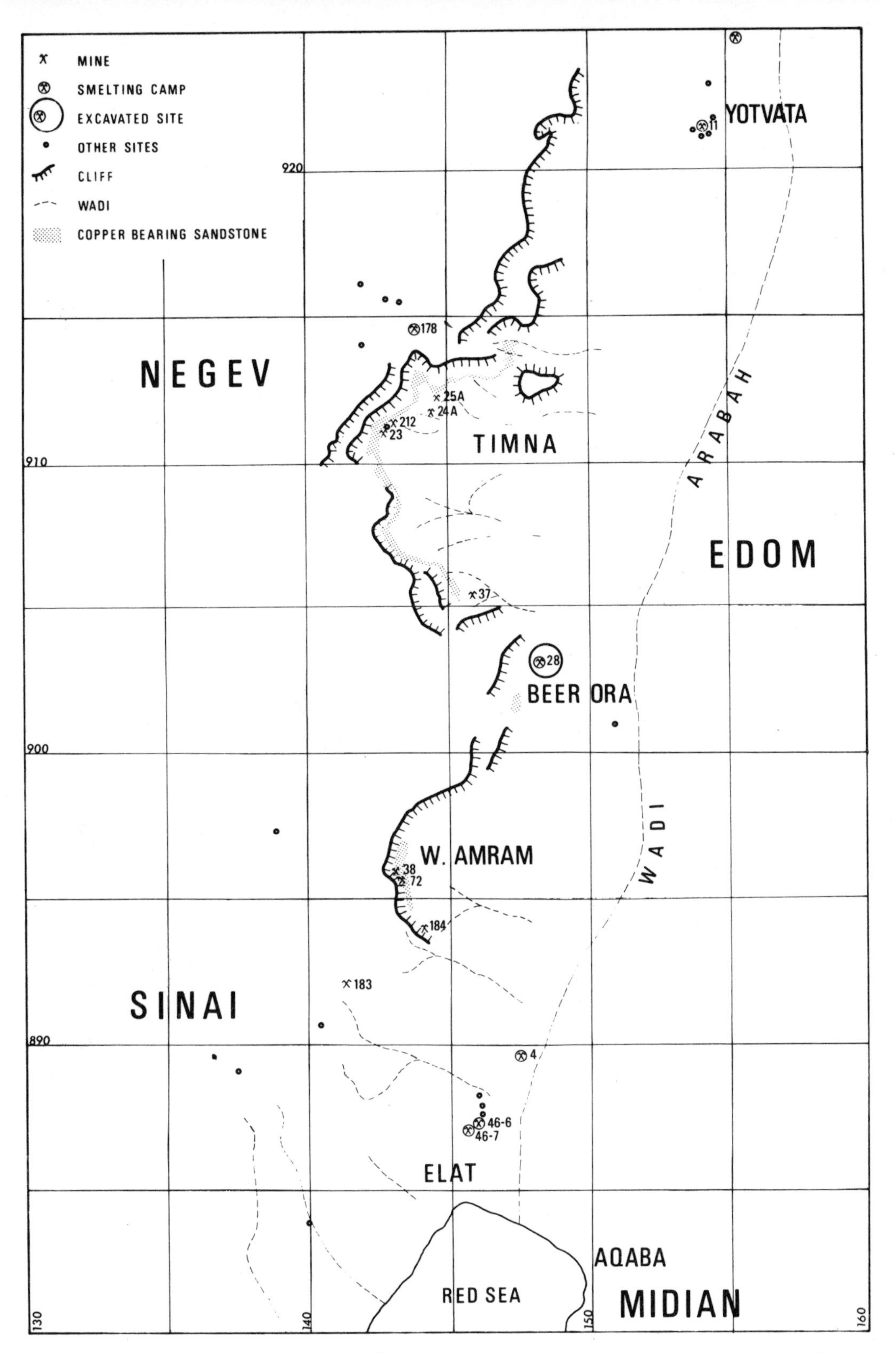

74 *Medieval sites in the Arabah and adjoining Negev mountains*

cutting of this blessed road'. In 1957 the author first investigated this section of the Darb el-Hagg and found an additional memorial inscription on top of the pass, built into the wall of the old Police Post. According to the late U. Ben-Chorin it reads: 'In the name of Allah the Compassionate and the Merciful this building was repaired in the name of the mighty, victorious Sultan Hasan and the mighty and victorious Sultan Mohammed Qul'aun on the 7th Ragab in the year 747' (=27th October, 1346). This is a new name and date to be added to the list of construction works undertaken along this part of the Darb el-Hagg.

The most difficult part of the descent through the steep cliffs involved cutting a wide, winding, solid road out of the rock. Here a large labour force must have been at work for a considerable length of time and it is not surprising, therefore, that a large workers' camp of the Mameluke period, Site 224 on the survey map of the Elot District (1968) at G.R. 13888882, was erected nearby. The location of this camp could not have been better chosen. It was placed on fairly even ground, hard to come by in this terrain, on a slope overlooking the beautiful rocky mountains of Elat, with the Red Sea and the Mountains of Edom and Midian in the background; obviously a fine place for a camp site and also for an archaeological excavation, undertaken by the Arabah Expedition in 1970.

The workers' camp, dated by pottery to the fourteenth century, consists of a large and widely spread-out group of well-built stone structures. Most of these structures were low circular enclosures, 3–5 m. in diameter, others were of irregular shape and had several rooms or compartments. On the lower part of the slope, two solid stone enclosures attracted special attention because of the stone structures in their centre, which gave the impression of being metallurgical furnaces. Ashes, charcoal and slaggy materials were found on top and around these structures and several heavy, slaggy, lumps of convex shape appeared to be parts of a furnace bottom.

Plate 125

Plate 127
Fig. 75

It was decided to excavate completely the larger of the two metallurgical enclosures. Measuring about 7×5 m. it was built as a semi-circular row of small compartments, open towards the east, and the original walls were still 1–1·5 m. high. Charcoal bits, some fragments of copper ore, much white wood ash and pieces of slag were found on the floor. There were also fragments of several iron tools and one complete iron nail. Although no complete working implements or any quantity of raw materials were found it nevertheless became apparent that this enclosure was a metallurgical workshop. The working tools, hammers and anvils, and all valuable

Fig. 77

materials must have been removed by the inhabitants upon leaving the site.

In the centre of the workshop stood a low stone structure, measuring 1 m. wide, 1·65 m. long and 0·60 m. high, solidly built of medium-sized, roughly dressed stones. One half of this structure was round, its inside full of small stones, with a large stone in its centre. The other half, almost exactly square, was a solid mass of stone which, podium-like, formed the base of a metallurgical furnace. There was a thick layer of ash and much unburned charcoal beneath which was the hard-burnt, cinder-like lining of a furnace bottom. This hearth had the shape of an elongated, oval, shallow crucible and consisted of an outer layer of clay against which a row of small flat stones, roughly 1 cm. thick, was placed in order to strengthen the furnace walls. The stone lining was protected against the fire by an additional clay layer 4 cm. in width, found burnt into a solid mass containing a lot of tiny ferruginous particles, perhaps hammer-scale. At the north end a partly attached, basin-like, stone structure was found filled with a quantity of clay. In the construction of the furnace clay was used as mortar between the courses of building stones, as furnace lining and also as a building material, especially on the south end of the furance. Near its north end a small aperture in in the furnace wall, penetrating through to the interior of the hearth,

Fig. 75

Plate 128

75 Site 224: Plan and section of a Mameluke smithy on the Darb el-Hagg, and a reconstruction of the furnace

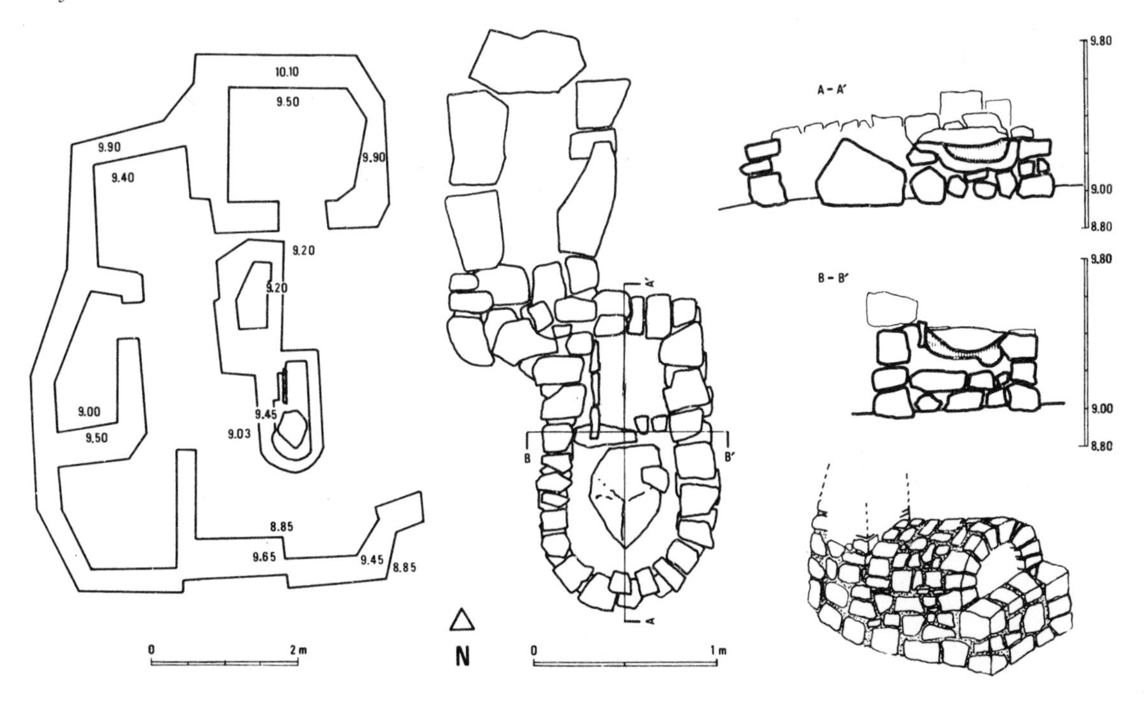

Fig. 76

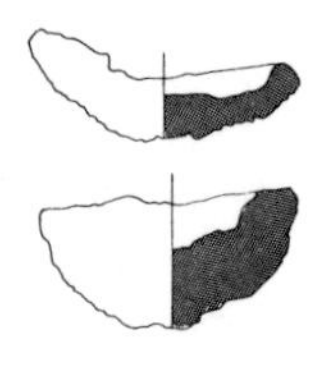

76 Smithing cinder from the furnace in Site 224

served as a tuyère for the bellows. Black ashes and charcoal were piled up on the floor next to the north end of the furnace, under a mass of stone debris from the furnace wall. In this pile a corroded socketed iron hoe was unearthed, which had obviously fallen out of the furnace. Several crucible-like, convex, slaggy lumps of cinder were also found in the debris containing, besides melted clay, a large quantity of unoxidized iron particles and scale.

The excavation of the furnace and the various finds made within it and on the floor around proved that the site must have been the workshop of a blacksmith of the Mameluke period. The following reconstruction of the furnace is proposed: the actual smithing furnace was a bellow-blown, bowl-type hearth, vaulted over by a semi-circular arched roof. Its opening was at the north side, with a thick layer of charcoal and ashes on the floor below it. Here, the partly-attached stone structure was a quenching tank or a water basin for the cooling of the smithing tools. A very similar smithy was found at Site 503 in the Wadi Zalaka, in South Sinai, with a complete stone tank *in situ*. At the south end of the smithing furnace a low chimney was built of small stones and clay to help ventilate the smithing furnace.

Fig. 75

More blacksmiths' workshops must have operated at Site 224, but much of the workers' camp was destroyed by later road construction and only furnace fragments, slaggy materials and cinders remained. There was no evidence for any copper smelting at the site although several pieces of copper ore were found in the workshop. The Mameluke smithy from Ras en Naqb is the only evidence for iron working in the area and represents the latest ancient phase of metallurgical activities found up to the present in the southern Arabah.

77 Fragments of iron tools found in the excavation of the Mameluke smithy (Site 224)

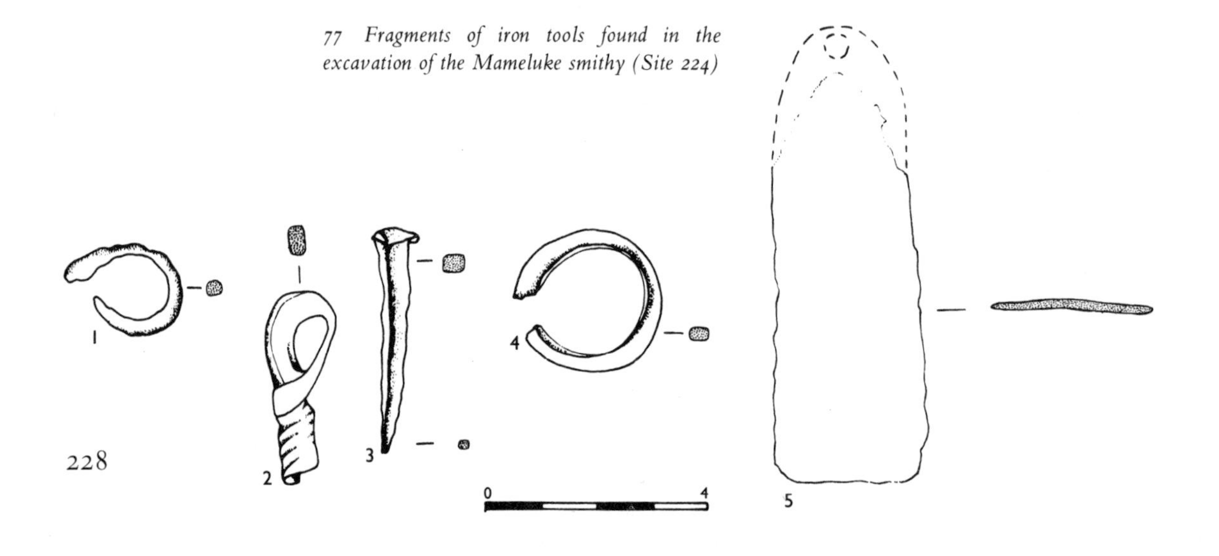

IX

6000 Years of Metallurgy in the Arabah

The Timna excavations uncovered remains of metallurgical activities dating from the Chalcolithic period to Mameluke times and proved that the Arabah was the scene of large-scale metal production and workings from the very beginnings of metallurgy. The modern Timna Copper Mines look back on more than 6000 years of metal-working in the Timna valley and the Arabah. This ancient tradition is reflected also in *Deuteronomy* (8:9): 'a land whose stones contain iron, and in whose hills you can mine copper'. Yet, there was never any real continuity in the exploitation of the ore deposits and the operation of smelting plants. Lapses of many generations, in fact, one of more than two thousand years separated the different periods of organized metallurgical activities in the area. Whenever the lands and deserts around Timna show traces of habitation, whether nomadic, sedentary or military or, alternatively, were utterly void, the copper mines show equal periods of operation or abandonment. It is evident that the periods of activity in the mines were altogether much shorter than the periods of abandonment and it is, therefore, rather astonishing to see that there exists, nevertheless, an almost continuous technological tradition in the metallurgical principles and installations of the Arabah.

Mining

The location of the mines, naturally a result of the position of suitable copper ores, was almost identical during the different periods of exploitation, but the extent of the mining and, naturally, its technology varied greatly from period to period. Chalcolithic mining consisted only of collecting the copper ore nodules which at that time must have been dispersed in considerable quantities all over the Middle White sandstone formations and in the wadi-beds below the Timna cliffs. The copper ore nodules were taken for 'dressing' to a working camp (Site 29), located on the edge of the Arabah, where the nodules were crushed on large heavy stone mortars and hand-picked to separate the ore from the unwanted siliceous gangue. Chalcolithic copper mining at Timna was a small enterprise, involving the actual work of no more than some tens of people.

The Late Bronze Age–Early Iron Age I, Egyptian–Midianite, copper mines were a large-scale enterprise. All Middle White formations of the Timna circus show evidence of intensive exploitation during this period. The vertical white sandstone walls were attacked with stone hammers and the copper nodules picked out of the heaps of shattered sandstone. By this method about 20 kg. of nodules could be mined by one man in a day's work, which is enough to produce 2–4 kg. of copper. 'Dressing' of copper and iron ore took place in the saucer-shaped hollows on the slopes next to the actual mining walls, where many grinding implements were found. Analysis (by A. Lupu) of the sand inside these hollows showed that it contained 0·1 to 0·25% copper and 5–12% iron while the sand outside contained only 0·005–0·01% copper and 2·1–3·5% iron. Why the dressing operation of copper ore nodules with such high copper content should have taken place here and not at the smelting sites, is a problem. Two alternative solutions to this problem may be proposed: the dressing operation could have been a winnowing process, utilizing the strong north wind prevailing on the wide open horizontal surfaces or, alternatively, the location of the operation near the mines was the result of the tribal organization of the labour force, *i.e.* the men were mining the ore and women and children were grinding it and sorting the ore from the gangue.

A completely different organization and technological system was introduced by the Romans using shaft mining and metal chisels to cut deep galleries into the mountain sides. The Roman mining was selective in so far as shafts were found only in areas of the Middle White Horizon especially rich in copper ore, and there numerous shafts were concentrated within the area of heaviest mineralization. Only some preliminary ore dressing took place near the mines where the broken sandstone often formed large dumps. The Roman smelters seem to have used shaft mining methods in the Middle White Horizon of the Arabah, where the copper ore nodules are dispersed throughout the white sandstone and do not appear in layers to be followed, because these were their traditional mining methods. The location of the individual shafts was often determined by the tendency of the ore nodules to form in relative concentrations along horizontally bedded fossilized trees and the Romans, with their metal tools, once a fossilized tree was located, would naturally follow such concentrations to the end. Nowhere in Timna, or at any of the other Arabah sites, did smelting take place at the mines or their immediate vicinity.

Ores and fluxes

Plate VIII

The copper ore appears mainly as cupriferous concretional nodules associated with fossilized trees. The texture of the wood in the fossilized remains is clearly visible and, according to J. Bentor and A. Slatkin, the organic material has been first substituted by pyrites which has then been partly or completely replaced by chalcocite. The nodules are varied in shape and composition; some are small, ball-shaped or oval, 3–5 cm. in diameter or length, others are larger lumps of irregular shape. Most of them are of complex composition consisting of oxidized sulphides with a varying degree of oxidization. Mineralogical examination of typical nodules showed a banded structure consisting of malachite, azurite, chalcocite and some cuprite and chrysocolla. The chemical composition of such typical complex nodules was found to be, according to A. Lupu (in %):

Cu	SiO_2	Fe_2O_3	MnO	CaO	MgO	Zn	Pb
25–37	20–30	4–6	0·1–5	0·1–0·3	0·1–2	traces	traces

Other samples, consisting of malachite and quartz, contained, according to P. Fields (in %):

Chemistry

Sample No.	SiO_2	CuO	R_2O_3	P_2O_5	Na_2O	K_2O	H_2O	CO_2
5A	17·82	58·90	0·79	<0·03	0·06	<0·04	7·87	13·13
			Fe_2O_3					
2A	33·75	36·55	1·04	<0·03	0·01	<0·02	3·53	15·13
	CaO	MgO	Al_2O_3	ZnO	SnO	MnO	BaO	
5A	0·08	0·15	0·15	1·0	0·02	0·05		
2A	6·06	0·16	0·29	0·37	—	0·03	<0·1	

Spectrochemical %

Sample No.	Ag	As	Sb	Co	Ni	Pb	Bi	Zn	Sn
5A	0·01	<0·05	<0·01	traces	traces	0·01	0·1	0·8	<0·01
2A	0·02	0·05	0·01	traces	traces	0·04	0·01	<0·3	<0·1

It seems that during the Chalcolithic beginnings of copper smelting the miners looked for nodules especially rich in copper and poor in silica. Copper ore found near the Chalcolithic smelting furnace at Site 39 was examined and consisted, according to A. Lupu, of (in %):

Cu	SiO_2	Al_2O_3	FeO	MnO	CaO	MgO
38·20–43·74	4·92–10·70	0·33–7·84	0·33–17·72	0–0·33	1·61–5·60	0·12–1·17

It is interesting to see that the Chalcolithic miners of Timna had already reached such a high degree of experience and differentiation.

Plate IX

The analyses of the slag from all periods of copper smelting in the Arabah, make it certain that fluxes were used in the smelting process to facilitate the separation of the metallic copper from the slag. The most suitable flux used, according to the analyses, from the Chalcolithic period onwards, was iron ore. It was obtainable from the Middle White Horizon of Timna in the form of fossilized wood which had been converted to haematite (60–85% Fe_2O_3 + 5–15% SiO_2). The Late Bronze Age–Early Iron Age I smelters also used manganese ore as flux which is obtainable in quantity in the Nahal Mangan, the northern part of the Timna Valley. Manganiferous concretions containing 30–80% MnO_2 and only 5–15% SiO_2 were actually found on the Ramesside smelting sites of Timna. Although analysis of some Chalcolithic slag showed a high manganese oxide content (29·33%) this should not be taken as conclusive evidence for the intentional use of manganese ore as flux at that period; it could have been used unintentionally together with iron oxide (30·35%) found in the same slag sample. No manganese flux was used in the Roman period.

There is evidence for the use by the Chalcolithic metalworkers of some material rich in lime as flux because some of their slag contains 8–23% CaO. The Egyptian–Midianite smelters must also have added lime flux to their smelting charge as their slag contains 2–14% CaO, which could not have come from the copper ore nodules (containing only 0·1–0·8% CaO). A very high lime flux component is present in all Roman slag (14–20% CaO) and must have been regularly added to the smelting charge. Although limestone is plentiful in the Timna area it is suggested that, at least in some cases, ground up shells were used as calcareous flux. This would explain why so many Triton shells, common in the Red Sea, were found on smelting sites of all periods at Timna as well as in Sinai.

Fuel

Charcoal made of local acacia trees and brushwood was used in all periods to reach a minimum temperature of 1100°C. Some dry donkey dung was found in the pits next to the smelting furnaces of Area G, in Site 2, and may also have been used as fuel. It is obvious that the large scale smelting activities in Timna and later at Beer Ora must have seriously depleted the flora of the Arabah.

The smelters' organization

The essential diversity in the locations of the smelting sites of the different periods of metallurgical activity in the Arabah reflects not only problems of supplies and security but also, to some degree, the social organization of the workers and the planning abilities of the ancient industrialists. The fact that complete metallurgical industries of periods some 6000 years apart were found in the Arabah in close proximity, made it possible to explore and compare their actual organization and to try to reconstruct their environmental conditions besides, of course, the highly informative study of the actual smelting installations and principles. The Chalcolithic copper workers were part of a large tribal community which had settled in the area of the copper deposits. Whilst the location of the mines and the dressing camps was a function of the ore deposits, the actual smelting sites were chosen because of their proximity to the acacia trees and sources of drinking water in the Arabah. The smelting installations were near habitation sites, often on hill tops where the prevailing north wind made life near the furnaces bearable, and the copper smelters seem to have had their living quarters within their tribal habitat. In fact, some of the smelting installations in the Arabah were found attached to habitation sites. All this points to a well-organized society, with specialization of work and distribution of responsibilities. The wide dispersion of the Chalcolithic habitation sites and copper works presents a picture of peace and safety and no defensive measures of any kind seem to have been necessary, although copper must have been extremely valuable in those days. The Chalcolithic copper production of the Arabah, though certainly of importance at the time and probably the *raison d'être* of the Chalcolithic settlement of the southern Arabah – and also of South Sinai – appears, nevertheless, still in a stage of transition from a nomadic, tribal home-industry to a properly organized industrial undertaking.

The Ramesside copper works present the picture of a large-scale industrial undertaking planned to meet the challenges of environment, climate, security and organization. The copper mines form a large semi-circle along the Timna Cliffs, with dressing camps on the slopes nearby. Drinking water was made conveniently available on the spot by deep cisterns cut into the mining walls, but all other supplies, including food and working tools, had to be brought from central supply stores located at the smelting sites. The mine workers appear to have had their living quarters in the central parts of the Timna valley where all the smelting sites were located. It is plausible to assume that the miners went to work in the early morning hours carrying all necessary supplies, and returned at night to their camp, everybody bringing along his daily quota of ore for the smelters. The habitation camps show conspicuous differences in the standard of construction and layout. Unlike the rather primitive camp sites 12 and 35, consisting of low stone enclosures spread out in more or less isolated family units which seemed to have served the miners and the primitive labour force, each with a small domestic fireplace, Site 3 showed rather well-built houses consisting of several rooms and must have served as a habitation site for the specialists, overseers or even the Egyptian overlords.

Habitation and smelting camps were located in a semi-circle around the west side of the Timna massif, at almost equal distances from the mines. It is assumed that there were good reasons for this choice of location which was not entirely the most convenient one if we remember that drinking water, fuel and food must have been brought to the sites from the direction of the Arabah. One of the main reasons must have been considerations of security and there is, at the north end of the industrial semi-circle, a watch tower (Site 17) and on the other end the two walled sites, 30 and 34. However, even if we assume that after some time the local tribes collaborated with the Egyptians in the copper works, the concentration of the sites deep inside the Timna valley and away from the Arabah seems, nevertheless, still mainly a result of security considerations. Ramesside Timna must have been a treasure island in an insecure environment at all times. On the other hand, the concentration of all working and habitation sites in a restricted area made the control over the workers and the distribution of supplies, raw materials and water, relatively easy tasks.

The Roman copper industry was planned according to entirely different considerations. Whilst the Roman shaft mines were situated in the Timna valley, Nahal Amram, Nahal Tuweiba and other locations, there was only one large Roman smelting centre in the Arabah, located near a good source of drinking water (Beer Ora), within easy reach of the trees and near the Roman road from Yotvata (ad Dianam?) to Aila. The widely dispersed scenes of Roman mining activities are a clear indication of the secure conditions prevailing in the area during this period. This highly centralized industry is assumed to have been run by the Third Legion *Cyrenaica*, perhaps with Christian forced labour. Its different parts were connected by a Roman road, and it seems clear that it was based on the complete military control of the area.

The latest, medieval, phase of ancient metalworking in the Arabah so far excavated belongs to a period of abandonment of the large-scale copper works of Timna. It consisted only of small-scale workings and was of a rather temporary nature.

The copper smelting processes and installations

The copper smelters of the Arabah used throughout the ages the same direct smelting process whereby metallic copper was produced from copper ores without any previous roasting process. The smelting charge consisted of finely-ground copper ores, carbonates and oxides, mixed in the correct proportions with equally finely-ground fluxes – iron or manganese oxides, limestone or sea shells – and charcoal. Possibly a very small amount of water was added to the charge to pelletize it and prevent the fines from blowing away. This charge was fed through the open top of the smelting furnace onto a charcoal fire where the ore was reduced to metallic copper and the other components of the charge turned to slag. The heavy metallic copper globules tended to sink through the fluid slag to the furnace bottom. The feeding of the smelting charge was repeated until the furnace was full. Some kind of ventilation appliances must have been used in all periods as the temperatures reached in the excavated furnaces were 1180–1350°C, which could not have been obtained without the use of an air blast.

Up to the smelting phase of the reduction of the copper ore into metallic copper globules there was little difference between Chalcolithic, Late Bronze Age or Byzantine copper smelting. From this phase on, however, there were considerable differences in technological refinement and efficiency between the different periods, although the zenith of metallurgical sophistication had been reached in the early Ramesside period, *i.e.* the Late Bronze Age. A criterion of metallurgical progress is the advancement in the technology of the smelting furnace and in the Arabah there existed two main stages of furnace development, represented by the earliest, primitive smelting bowl-hearth and the advanced and sophisticated tapping furnaces.

The Chalcolithic smelting furnace was essentially a hole in the ground, with a stone-built upper part standing above ground. There was no furnace lining and no tapping arrangement. Some kind of ventilation must have been provided although no archaeological evidence as to its nature was found. After the furnace became filled up with smelted material, the slags and copper were allowed to cool inside. To extract the copper meant, therefore, the breaking of the upper furnace walls. Some of the metallic copper had sunk down onto the furnace bottom but very much of the copper remained in the slag. In fact, the smelting slag was found to contain 5–15% copper which is about half of the total copper produced. This slag-copper mixture was apparently separated by breaking the slag and picking out the metallic copper drops. Metallic copper globules found in the slag from the Chalcolithic smelting furnace contained (A. Lupu): 97·43–98·88% copper and 0·24–0·30% iron, some cuprous oxide and small quantities of lead.

The Ramesside smelting furnace was also a bowl furnace but it was built into a specially excavated hollow together with a stone-flanked tapping-pit and was provided with tuyères for bellows. Two bellows seem to have been in action at the Late Bronze–Early Iron Age furnace, one penetrating diagonally through the wall at the back of the furnace, a second going horizontally through the front wall, above the tapping pit. The furnace walls and bottom were lined with a thick layer of very calcareous mortar (up to 40% CaO). As soon as the smelting bowl became filled with smelted materials, the bellows were stopped and a tapping hole was drilled through the lower part of the furnace front wall, above the slag pit. Care was taken not to drill this hole too low in order to keep the metallic copper, which had collected on the furnace bottom, inside the furnace, but allowing most of the fluid slag to run off into the slag pit. The bun-shaped copper ingot, formed on the furnace bottom, was quickly removed, perhaps by plunging a copper rod into the just-molten copper, leaving it until the ingot solidified, and then pulling both together out of the furnace. The tapped slag also solidified in its slag pit and was immediately thrown onto the adjoining slag heap by means of a metal hook catching hold in the cast-in hole in the slag circle. These sophisticated arrangements seem to have allowed a continuous smelting operation, thereby conserving much fuel. As soon as the tapped slag and copper was removed, any slag still remaining inside the furnace was also quickly taken out. The furnace would then be fettled and the slagged clay protectors of the bellows ends exchanged for new ones. All this could be done fairly fast and a new charge could be fed into the still hot furnace. Some of the furnace slag of the light porous type, which was similar to the Chalcolithic slag, contained many copper globules and was broken up to recover the metal. These copper globules would then be melted in the nearby crucible melting furnaces, perhaps to make ingots or votive gifts for the Hathor temple.

In one smelting furnace with the capacity of between 70 and 100 litres, 200–400 kg. of smelting charge could be smelted at one time, producing between 20–60 kg. of copper per day. The copper contains (according to A. Lupu): 89–92% Cu, 5·0–9·7% Fe, 0·2–1·2% Pb, 0·01–0·1% S. Spectrochemically established further elements are (P. Fields): 0·06% Ag, 0·06–0·07% As, 0·01% Sb, 0·1–0·3% Ni, 0·2–0·3% Co, 0·8–0·1% Zn, tr. 0·05% Sn, 0·01% Bi.

The relatively high lead content of the metallic copper from Timna was already noticed by the late B. H. Mcleod, who wrote in his appendix to the present author's first Arabah report (in the *Palestine Exploration Quarterly*, 1962): 'The copper, however, contained more lead in proportion to the copper than was found in the ore'. This problem remained unsolved until the analytical results of the Timna samples became available in 1966–8 and it was established that the lead content must have derived from the fluxes. In one flux sample as much as 10·5% lead was found. These facts are of decisive significance for the problems concerned with the tracing of metal objects back to their sources by comparative spectrographic analyses of metal objects and mineral deposits. In the light of the results of the analytical work done on complete sets of related samples from Timna, from the mines and through all processes to the finished metal objects, it seems most problematic to try any tracing back to ore sources by comparing element patterns of metals and ores without detailed extractive metallurgy. It now became clear that the fluxes are just as important in the formation of element patterns in metal as the ores and the production processes, and these results cast serious doubts, to say the least, on the practicability and usefulness of traces and/or major element patterns as a means of relating metal objects to their ore sources.

The Roman smelting furnace was, in principle, the same as the Late Bronze–Early Iron Age furnace, except that the smelting bowl was a separate pit dug in the ground, its rim strengthened by a semicircle of stones, with the tapping pit dug as a separate, shallow hollow next to it. There were no flanking stones to protect the slag pit and only one bellows seems to have been used. The furnace was lined with non-calcareous clay which became very hard after firing. The tapping hole was low down near the furnace bottom and at least one case was found of metallic copper having escaped from the furnace together with the slag. The slag circles showed the same cast-in techniques of the centre hole as found in the Late Bronze Age smelting sites.

Metallic copper from Beer Ora contained (A. Lupu): 94·48% Cu, 2·88% Fe (3·64% FeO), 0·30% Pb. Elements found spectrochemically were (P. Fields): 0·02% Ag, 0·08% As, 0·01% Sb, 0·08% Ni, 0·07% Co, 0·5% Pb, 0·2% Zn, traces of Sn, <0·01% Bi.

Roman copper smelting technology, though well advanced, did not show any essential improvements, compared with the end of the second millennium BC. The same furnaces were also used in post-Roman times all over the Near East without any change of design and the same furnace type was still used for iron smelting in Roman and medieval Europe. No smelting furnace of medieval Arab times has yet been excavated in the Arabah but smelting slag of the tapped-type found on two sites of this period (sites 46/2 and 64) indicated that the Arabic smelting furnace continued the Roman smelting traditions.

Crucible furnaces and the manufacturing of metal objects

There is no evidence in the Arabah from the Chalcolithic period of the methods or installations used to melt copper. It seems possible that the copper globules, mechanically separated from the slag-copper mixture, were melted together in a crucible, as Chalcolithic crucibles for melting copper were found at Abu Matar near Beer Sheba. But it is also possible that small copper pieces were forged from these globules by cold or hot hammering. A small copper awl found during the survey in a Chalcolithic site (375) was made of impure copper containing stringers of slag and curpous oxide. It had been mainly hot worked, with some finishing cold work, and the grain size of the metal was small and the hardness (HV5) 83. (R. F. Tylecote). The metallurgical cycle of Late Bronze Age to Byzantine Timna was completed with the making of finished copper tools. Unfortunately, Timna was only a primary producing area, where copper was produced but not worked into implements on any scale. The melting furnaces were either used to cast the copper globules extracted from the furnace slag into ingots, or to make tools for the local use of the workers. There was, however, no industrial or commercial casting of tools in Timna at any time. The crucible furnaces were but simple heating devices. Copper pieces were placed in crucibles inside the furnace and charcoal piled all around them to prevent excessive oxidation of the melted copper and to obtain the required temperature of about 1200°C. The thick-walled clay crucibles and the loss of heat in the simple open furnaces, made such a high temperature necessary. Small clay tuyères for the crucible furnaces were found at Ramesside Timna but none were found at Beer Ora, although bellows must have been used there also.

Copper implements from the Early Iron Age I, Site 2, were metallographically investigated by R. F. Tylecote and are clearly of local manufacture. A spear-butt consisted of impure copper which had been hot-worked or cold-worked and annealed. This was followed by some cold work to harden the surface of the tool. Its hardness was 107 HV5 (compared with 40HV of 'as cast' pure copper). A toggle pin, made of cast copper and forged, could not be measured for hardness because of corrosion. A piece of thin sheet metal showed hardness of 60 HV5. It seems to be a piece of impure copper which has been worked and finally annealed to reduce its hardness. Other implements showed similar properties. The tip of an awl showed evidence of heavy use, its hardness at the surface being 151 HV5.

The manufacturing methods for implements in Ramesside Timna were simple, even primitive, as should be expected from an industry mainly concerned with extractive, primary production, and employing local, semi-nomadic workers. The copper, after being melted in crucibles, was cast into sheets or small bars. Numerous castings of such shapes were also found in the temple. These were the rough products of the small casting workshops and were subsequently made into objects of required shapes by hot or cold forging. However, in the Hathor Temple, besides hundreds of rings, armlets, jewellery and make-up equipment, primitively shaped by forging, finely cast and finished figurines, as well as gilded copper and iron objects, were found representing an advanced stage of metalworking. These non-Egyptian metal objects were apparently made by the Kenites–Midianites and testify to a long-standing tradition of fine metalwork in the Arabah and its neighbourhood. Like the smelting technology of the Roman and medieval Arabic periods, the methods of toolmaking seem to have carried on Bronze Age traditions of metallurgy and metalworking. Future excavations will have to discover whether the same traditions were still operative from Late Arab to recent times.

Bibliography

Aharoni, Y. *The Land of the Bible*, Philadelphia, 1967.

— *The Iron Age Pottery from Timna* (in the press)

Alt, A. Aus der Araba, II–IV, *Zeit. Deutschen Palästina-Vereins* 58 (1935).

Bartura, Y. *Type Sections of Paleozoic Formations in the Timna Area*, Israel, Ministry of Development, G.S.I., 1966.

Breasted, J. H. *Ancient Records of Egypt*, Chicago, 1906–7, vol. IV, 204.

Ferembach, D. and Dotan, M. A propos d'un crane trépané trouvé a Timna, *Bull. Soc. d'Anthropol. de Paris*, 1957, 10 series, 245–75.

Frank, F. Aus der Araba, I, *Zeit. Deutschen Palästina-Vereins* 1934.

Giveon, R. The Shosu of Egyptian Sources and the Exodus, *Fourth World Congress of Jewish Studies, 1967*, vol. I, 193–6.

— *Les Bédouins Shosou des Documents Égyptiens*, Leiden, 1971.

Glueck, N. Explorations in Eastern Palestine, II, *Bull. American Schools of Oriental Research* XV, 1935.

— *The Other Side of the Jordan*, New Haven, Conn., 1940.

— *Rivers in the Desert*, London and New York, 1959.

— Ezion-geber, *Biblical Archaeologist* 1965, 70–87.

Kind, H. D. Antike Kupfergewinnung zwischen Rotem und Totem Meer, *Zeit. Deutschen Palästina-Vereins* 81 (1965), 56–73.

Kitchen, K. A. Some New Light on the Asiatic Wars of Ramesses II, *J. Egypt. Arch.* 50 (1964), 47–70.

Lupu, A. Metallurgical aspects of Chalcolithic copper working at Timna (Israel), *Bull. Historical Met. Group*, 1970, 21–3.

Mcleod, Bentley H. The Metallurgy of King Solomon's Copper Smelters, *Palestine Exploration Quarterly* 1962, 68–71.

Musil, A. *Arabia Petraea* II, Vienna, 1907.

Parr, P. J., Harding, G. L. and Dayton, J. E. Preliminary Survey in N.W. Arabia, 1968, *Bull. Inst. of Archaeology* Nos 8–9, London, 1970, 193–242.

PERROT, J. Excavations at Tell Abu Matar, *Israel Exploration J.* 5 (1955).

— Palestine–Syria–Cilicia, in *Courses towards Urban Life*, 1962, 147–64.

PETHERICK, J. *Egypt, the Soudan and Central Afrika*, . . . Edinburgh and London, 1861.

PETRIE, W. M. F. *Researches in Sinai*, London, 1906.

ROTHENBERG, B. King Solomon's Mines, A new discovery, *Illus. London News* Sept. 3, 1960.

— *God's Wilderness, Discoveries in Sinai.* London, 1961.

— Ancient Copper Industries in the Western Arabah, *Palestine Exploration Quarterly* 1962, 5–71.

— König Salomons Hafen im Roten Meer neu entdeckt, *Das Heilige Land*, Bonn, 1965, pp. 19–28.

— Excavations in the Early Iron Age Copper Industry at Timna, 1964, *Zeit. Deutschen Palästina-Vereins* 1966, 125–35.

— The Chalcolithic Copper Industry at Timna, *Bull. Mus. Haaretz* 8, 1966, 86–93.

— and LUPU, A. Excavations at Timna, 1964–1966, *Bull. Museum Haaretz*, 1967.

— *Negeb, Archaeology in the Negeb and the Arabah* (in Hebrew) 1967.

— and COHEN, E. An Archaeological Survey of the Eloth District and the Southernmost Negev, *Bull. Mus. Haaretz* 10, 1968, 25–35.

— King Solomon's Mines no more, *Illus. London News* Nov. 15, 1969.

— The Egyptian Temple at Timna, *Illus. London News* Nov. 29, 1969.

— Un temple égyptien decouvert dans l'Arabah, *Bible et Terre Sainte* 123, 1970.

— An Egyptian Temple of Hathor discovered in the Southern Arabah (Israel), *Bull. Museum Haaretz*, 12, 1970, 28–35.

— An Archaeological Survey of South Sinai, *Palestine Exploration Quarterly* 1970, 4–29; also *Bull. Museum Haaretz*, 11, 1969, 22–38.

— and LUPU, A. The Extractive Metallurgy of the Early Iron Age Copper Industry in the Arabah, Israel. *Arch. Austriaca* 1970, 91–130.

— The Arabah in Roman and Byzantine Times in the light of new research, *Roman Frontier Studies 1967*, Tel Aviv, 1971.

SLATKIN, A. Nodules Cuprifères du Neguev Meridional (Israel), in *Bull. Research Council of Israel*, Section G, Vol. 10G, 1961, 292–301.

— *Etude Petrographique Comparée de Certaines Poteries Anciennes D'Israel* (in the press).

TYLECOTE, R. F., LUPU, A. and ROTHENBERG, B. A Study of Early Copper Smelting and Working Sites in Israel, *J. Inst. of Metals* 1967, 235–43.

— A metallurgical investigation of material from early copper working sites in the Arabah, *Bull. Histor. Metal. Group*. 1968, 86–8.

— *Metallurgy in Archaeology*, London, 1962.

WURZBURGER, U. Copper Silicates in the Timna Ore Deposits, *Israel Journal of Chemistry* 8, 1970, 443–57.

YADIN, Y. The Earliest Record of Egypt's Military Penetration into Asia? *Israel Exploration J.* 5, (1955).

List of New Hebrew and Old Arabic Place Names

Beer Metek	Bir Metek
Beer Milhan	Bir el Milhan
Beer Ora	Bir Hindis
Biq 'at Sayarim	Wadi Jaradi
Borot Roded	Thamilat er-Radadi
Biq 'at Uvda	Wadi Ijfi
Har	Mountain
Har Haargaman	Jebel Ghadyan
Jeruham	Tell Rakhma
Makhtesh Ramon	Wadi Raman
Malehat Yotvata	Sibkh et Taba
Mezad Gozal	Rujm Umm Zoghal
Nahal	Wadi ('N' is today officially used for 'Wadi' in Israel)
N. Amram	Wadi Amrani
N. Mangan	Wadi Barak
N. Nehushtan	Wadi Umm Ghadhak
N. Nimra	W. Numra
N. Quleb	Wadi Quleb
N. Raham	Wadi Rakhama
N. Roded	Wadi Radadi
N. Timna	Wadi Atshana
Timna	Munei 'iya (Mene 'iyeh)
Yotvata	Ain Ghadyan
Zuge Timna (or Timna Cliffs)	Jebel el Munei 'iya

List of Illustrations

The site photographs were taken by the author and the objects by Abe Hai. All photographs are published by courtesy of the Arabah Expedition.

Colour Plates

Monochrome Plates

Figures

Index

Index